纺织高等教育"十一五"部委级规划教材
纺织工程专业双语教材

Introduction to Textiles
纺 织 导 论

顾平　主编

中国纺织出版社

内 容 提 要

本书主要介绍纺织纤维、纺织纱线、织物成形系统(机织、针织、非织造及其他　形方法)、织物染整以及纺织品测试与保养等内容,同时阐述了纤维、纱线、织物的结构、性能及其基本生产工艺等知识。

本书可作为高等纺织院校纺织、服装及相关专业学生的双语教材或专业英语教材,也可供从事纺织、服装贸易工作的相关人员使用,同时可作为纺织企业的英语培训教材。

Abstract

This book mainly introduces the forming systems of textile fibers, textile yarns, fabric (including weaving, knitting, nonweaving and other construction processing), as well as textile dyeing & finishing, textile testing & care. It also deals with the basic structures and properties of fibers, yarns, fabrics as well as the fundamental manufacturing techenology of these products.

The authors hope this book will be helpful for those students who major in Textile, Fashion Clothing or other curricula in colleges and universities, for persons who engage in textile, fashion trade. It can be used as a bilingual or training textbook.

图书在版编目(CIP)数据

纺织导论/顾平主编.—北京:中国纺织出版社,2008. 11(2023. 1 重印)

纺织高等教育“十一五”部委级规划教材

纺织工程专业双语教材

ISBN 978-7-5064-5293-9

Ⅰ.纺…　Ⅱ.顾…　Ⅲ.纺织—双语教学—高等学校—教材　Ⅳ.TS1

中国版本图书馆 CIP 数据核字(2008)第 132719 号

策划编辑:张冬霞　　责任编辑:曹昌虹　　特约编辑:张烛微

责任校对:陈　红　　责任设计:李　然　　责任印制:何　建

中国纺织出版社出版发行

地址:北京市朝阳区百子湾东里 A407 号楼　邮政编码:100124

销售电话:010—67004422　传真:010—87155801

http://www. c-textilep. com

中国纺织出版社天猫旗舰店

官方微博 http://weibo. com/2119887771

北京虎彩文化传播有限公司印刷　各地新华书店经销

2008 年 11 月第 1 版　2023 年 1 月第 4 次印刷

开本:787×1092　1/16　印张:14. 25

字数:300 千字　定价:38. 00 元

出版者的话

全面推进素质教育，着力培养基础扎实、知识面宽、能力强、素质高的人才，已成为当今本科教育的主题。教材建设作为教学的重要组成部分，如何适应新形势下我国教学改革要求，与时俱进，编写出高质量的教材，在人才培养中发挥作用，成为院校和出版人共同努力的目标。2005 年 1 月，教育部颁发了教高[2005]1 号文件"教育部关于印发《关于进一步加强高等学校本科教学工作的若干意见》"（以下简称《意见》），明确指出我国本科教学工作要着眼于国家现代化建设和人的全面发展需要，着力提高大学生的学习能力、实践能力和创新能力。《意见》提出要推进课程改革，不断优化学科专业结构，加强新设置专业建设和管理，把拓宽专业口径与灵活设置专业方向有机结合。要继续推进课程体系、教学内容、教学方法和手段的改革，构建新的课程结构，加大选修课程开设比例，积极推进弹性学习制度建设。要切实改变课堂讲授所占学时过多的状况，为学生提供更多的自主学习的时间和空间。大力加强实践教学，切实提高大学生的实践能力。区别不同学科对实践教学的要求，合理制定实践教学方案，完善实践教学体系。《意见》强调要加强教材建设，大力锤炼精品教材，并把精品教材作为教材选用的主要目标。对发展迅速和应用性强的课程，要不断更新教材内容，积极开发新教材，并使高质量的新版教材成为教材选用的主体。

随着《意见》出台，教育部组织制订了普通高等教育"十一五"国家级教材规划，并于 2006 年 8 月 10 日正式下发了教材规划，确定了 9716 种"十一五"国家级教材规划选题，我社共有 103 种教材被纳入国家级教材规划。在此基础上，中国纺织服装教育学会与我社共同组织各院校制订出"十一五"部委级教材规划。为在"十一五"期间切实做好国家级及部委级本科教材的出版工作，我社主动进行了教材创新型模式的深入策划，力求使教材出版与教学改革和课程建设发展相适应，充分体现教材的适用性、科学性、系统性和新颖性，使教材内容具有以下三个特点：

（1）围绕一个核心——育人目标。根据教育规律和课程设置特点，从提高学生分析问题、解决问题的能力入手，教材附有课程设置指导，并于章首介绍本章知识点、重点、难点及专业技能，增加相关学科的最新研究理论、研究热点或历史背景，章后附形式多样的思考题等，提高教材的可读性，增加学生学习兴趣和自学能力，提升学生科技素养和人文素养。

（2）突出一个环节——实践环节。教材出版突出应用性学科的特点，注重理论与生产实践的结合，有针对性地设置教材内容，增加实践、实验内容。

（3）实现一个立体——多媒体教材资源包。充分利用现代教育技术手段，将授课知识点制作成教学课件，以直观的形式、丰富的表达充分展现教学内容。

教材出版是教育发展中的重要组成部分，为出版高质量的教材，出版社严格甄选作者，组织专家评审，并对出版全过程进行过程跟踪，及时了解教材编写进度、编写质量，力求做到作者权威，编辑专业，审读严格，精品出版。我们愿与院校一起，共同探讨、完善教材出版，不断推出精品教材，以适应我国高等教育的发展要求。

中国纺织出版社

教材出版中心

前　言

《纺织导论》是为双语教学编写的，介绍了纺织领域的基础知识和最新进展，内容涵盖纺织纤维、纱线、织物成形、织物染整、纺织品测试与保养等方面的基本知识和原理以及各道加工制品的基本性能和总体发展趋向。读者可从中全面了解纺织这一富有活力的综合性行业。

本教材主要参考了美国及英国原版纺织教科书的内容（见参考文献），章节的编排按照纺织品的加工顺序展开，并结合编者多年的教学经验，同时兼顾学生的学习兴趣。

全书由九章构成：第一章介绍纺织工业、纺织教育方面的知识；第二章讨论常用天然纤维和化学纤维的生产加工、形态结构与主要性能；第三章介绍短纤维环锭纺纱工艺、新型纺纱方法、长丝纱的加工、纱线的类别和基本品质指标；第四到第七章涉及织物的成形加工，详细说明机织、针织、非织造三种成形工艺与设备以及三种织物的组织结构与形态，并简述织物的其他成形方法，如编织、钩织、簇绒和层压法等；第八章简要阐述纺织品前处理、染色、印花和整理的基本原理、加工方法和工艺，并介绍常用的染料。第九章包含纺织品常规测试指标、测试方法和纺织品保养方法及使用须知等。

《纺织导论》是纺织专业的入门课程，开始学习时读者并不需要具备纺织背景知识。为此，书中对技术术语和专业词汇的解释力求清楚，简明易懂；对有关的工艺理论、概念和方法的说明也力求深入浅出、结合实际，便于阅读理解。本书的编写特色如下：

(1)图文并茂。说明性的图表使复杂的工艺清楚易懂，有助于读者加深理解。归纳性的图表对所列各项内容的比较，有利于读者快速地把握主要知识点。

(2)前后文呼应。章节内容的编排兼顾了本书既作为教科书，又作为参考资料的需求。尽管各个专题相互独立，但前后文呼应、脉络清晰，使读者能获取更有条理的专业信息。

(3)章后有思考题和实训题。思考题供课堂学习讨论和复习，实训题为建议学生进行的课外活动内容。两者均联系实际，加深读者对本书内容的理解。

(4)专业词汇的中文标注。新出现的专业词汇、术语有中文注释，使读者能快速地理解其在课文中的含义。

(5)词汇表。教材附带的光盘中，列出了纺织纤维、纱线和织物（机织、针织和非织造）及工艺和相关制品的词汇，并选择了部分正文中出现的生词，对其基本含义做了解释。

(6)多媒体教学。本书附带光盘，光盘中包含两个附录和纺织词汇表以及正文中基础内容的多媒体教材，由 Mr. David J. Hogness（美国）朗读配音。

本教材由顾平主编。其中：第一章、第四章和第七章由顾平编写，第二章和第九章由杨旭红编写，第三章由马建兴编写，第五章由林红编写，第六章由王家俊编写，第八章由邢铁玲编写。编者除浙江理工大学的王家俊外，其他均来自苏州大学。全书最后由顾平统稿，天津工业大学黄故主审。

欢迎广大读者对本书提出宝贵意见和建议。

编者

2008 年 8 月

课程设置指导

本课程设置意义 《纺织导论》作为纺织工程、针织、纺织材料与纺织品设计、非织造材料和工程、服装、染整等专业的专业课程，囊括了从纺织纤维、纱线、坯织物到成品织物以及纺织品测试与保养等基本知识、基本工艺原理以及各道加工制品的基本性能和总体发展趋向。《纺织导论》课程在大学一年级开设，目的是使学生刚入学就能比较系统地了解纺织概况，明确纺织专业生产与纺织品开发对从业者知识结构、技能的要求，积极主动地学习大学阶段设置的各门基础课程和专业课程。《纺织导论》采用双语教学，对提高学生纺织专业英语的阅读与理解能力，积极高效地阅读国外纺织文献、开阔视野、拓宽知识面都有帮助，可为日后专业课的学习打下良好的英文基础。

本课程教学建议 《纺织导论》课程作为纺织工程(包括针织、纺织材料与纺织品设计、非织造材料)等专业课，建议理论授课45课时，每课时讲授字数控制在4000字以内，教学内容包括本书的全部或80%内容。

服装、染整等专业可作为选修课程，建议理论授课36课时，每课时讲授字数控制在4000字以内，选择与专业有关的内容讲授。

本课程教学目的 通过本课程的学习，学生应掌握常用纺织纤维的生产、形态与性能，短纤维环锭纺纱的工艺原理、新型纺纱方法、纱线的类别和基本品质指标，机织物、针织物、非织造布及其他织物的成形方法、织物组织、结构和形态，纺织品前处理、染色、印花和整理加工的基本原理、加工方法和工艺，纺织品常规测试指标、方法和保养要求等基本知识，为学习后续专业主干课程起引导作用；同时为学生掌握相应的纺织英语词汇、熟悉专业语言的英语表述方法、顺畅阅读和正确翻译专业文献奠定坚实基础。

Contents / 目录

Chapter 1 General Introduction
绪论

Textiles have such an important bearing on our daily lives that everyone needs to know something about them. From earliest times, people have used textiles of various types for covering, warmth, personal adornment, and even to display personal wealth. Today, textiles are still used for these purposes and everyone is an ultimate consumer. Included among consumers are merchandisers of many types, from the wholesale textile manufacturer and merchant to the sales force in any retail-store. Many industries, such as the automobile industry, are important consumers of textiles in various forms. Some other consumers are homemakers, dressmakers, interior decorators, and retail-store customers, as well as students who are studying for these and various other occupations and professions in which knowledge of textiles is of major importance.

装饰品
消费者/商品
批发商/商人/零售店

The merchant, particularly, and all those engaged in the purchase and/or sale of goods must be thoroughly familiar with the merchandise they are handling if they wish to be successful. In fact, the only way to achieve this ability is through knowledgeable awareness of technical data and an understanding of how to use these data. Knowledge of chemical, physical, and microbiological properties of fibers and of the behavior that results from techniques used in creating fabrics guides the consumer in making wise selections when purchasing textile products. Knowledge of proper maintenance techniques for the many fibers, fabrics, and finishes helps ensure satisfaction after purchase. These are the areas of knowledge that this book explores. This book, then, attempts to provide key ideas and facts that will help readers become truly informed consumers. These facts form the basis for the study of textile science.

保养
购买

1.1 Introduction 引言

1.1.1 Reasons for Studying Textiles 学习纺织品的原因

Textile science covers a large area. To begin with, the word textile comes from the Latin textilis, "woven", which in turn comes from the Latin verb textere, "to weave". In textile science, however, a textile is freely defined as any product made from fibers; thus the term refers not only to woven fabrics, but also to knitted fabrics, nonwoven fabrics, and special fabric constructions. The term textile fiber refers to any product capable of being woven or otherwise made into fabrics. These

机织物/针织物
非织造布/纺织纤维

broad reciprocal definitions are now accepted and used by nearly everyone who works with fibers or fabrics at any stage of manufacture or processing.

衣服/家具装饰 家用纺织品 过滤布

Textile fibers and the fabrics made from them have almost limitless uses, and new applications are constantly being found. The uses of fibers in clothing, home furnishings, and household textiles are familiar to everyone. But fibers are also used in the building trades and as insulation in appliances; they are used by industry for such products as filter cloths, pulley belts, and conveyor belts; they are used in all forms of transportation, including space exploration. In fact, fibers play their essential roles in nearly every type of activity or situation conceivable, from playing in the sand to walking on the moon.

原材料/成品

There is a logical development of raw material into finished consumers' goods. Studying textile materials in the interesting sequence of "fiber to yarn to fabric" will help you understand the construction and ultimate qualities of the fabrics with which you will become familiar. Here are the steps in the manufacture of fabrics from raw material to finished goods:

纤维/纺纱/加捻

(1) Fiber – raw material, which is either spun (or twisted) into yarn or else directly compressed into fabric.

纱线

(2) Yarn – arrangement of fibers, which is woven, knitted, or otherwise made into fabric.

织物/整理加工 消费品

(3) Fabric – arrangement of yarns or fibers, which by various finishing processes becomes finished consumers' goods.

A study of textiles will show, for example, why certain fabrics are more durable and therefore more serviceable for specific purposes. It will explain why certain fabrics make cool wearing apparel as well as give an impression of coolness when used as decoration. The matter of cleanliness and maintenance must also be estimated before purchasing, when that is an important factor.

品牌

Complete knowledge of textiles will facilitate an intelligent appraisal of standards and brands of merchandise and will develop the ability to distinguish quality in fabrics and, in turn, to appreciate the proper uses for the different qualities. As a result, both the consumer merchant and customer will know how to buy and what to buy, and salespeople will know how to render good service to those consumers who have not had the advantage of a formal course in textiles.

混纺 注册商标

Great strides have been made in the textile industry, and have markedly influenced our general economic growth. The prosperity and growth of related industries, such as petroleum and chemistry, and dependent industries, such as retail apparel stores, have produced broader employment opportunities. Competition for the consumer's dollar has fostered the creation of new textile fibers with specific qualities to compete with well-established fibers. New fiber blends have been created to combine many of these qualities into new types of yarns with new trademarks. There are also new names for the fabrics made of these new fibers and yarns. New finishes have been developed to add new and interesting characteristics to fibers, yarns, and fabrics.

This welter of creativity and the myriad of trademarks present a challenge to the consumer, who is sometimes knowledgeable but frequently confused. Without being overly technical, this information can be easily understood and consequently very useful to the consumer in business and personal life. All of this information can be adopted for such utilitarian benefits as economy, durability, serviceability, and comfort, as well as for such aesthetic values as hand (or feel), texture, design and color. 手感/质地/花型

In the study of textiles, the students' initial interest will become an absorbing interest when they discover the natural fascination of fabrics and their cultural associations, particularly when factual study is supplemented by actual handling of the textile materials. The subject will seem worthwhile as they become familiar with illustrative specimens and fabrics and begin to handle and learn to compare the raw materials of which fabrics are made as well as the finished consumers' goods. 样品

1.1.2 Ranges and Markets of Textile Industry 纺织工业的领域与市场

The textile industry is very complex. At this point, only an introduction can be given so as to provide a brief overview for some insight. It begins in agriculture with fiber production of cotton, flax, and other fibrous plants; in husbandry of sheep, other animals and silkworms; in mining of metals and minerals; in forestry for wood; in chemical research and production of synthetics. These fibers may be short or long, fine or coarse, soft or stiff, smooth or rough, and are processed into yarns and/or fabrics. Usually, however, the fibers are initially formed into yarns. These yarns may be simple or complex, single or ply, smooth or rough, highly twisted or loosely twisted. The yarns are then made into fabrics by various methods, such as weaving, knitting, nonwoven processing, braiding, or knotting. 棉/亚麻/牧羊业 蚕 强捻 弱捻 编织/打结,结网

Selected finishing or "wet-processing" procedures, the fabrics are converted into finished cloths, which provide particular appearances and performances. These fabrics are made into end-use products, including apparel, home furnishings, and various industrial applications. These products are then merchandised and sold. Every one of these aspects of the textile industry is a field in itself, and there is an interdependency with multiplying effects on other industries. 湿加工 外观 终端产品/服装 工业用途

As one learns about the various aspects of the production of textiles and textile products, it becomes apparent that these activities play a major role in the economy. The industry has so many facets that it is possible that the student's interests may develop in the pursuit of a career in one of them. For example, consumers require the services of teachers of textiles, home economics, and interior decoration. 室内装潢

Opportunities for careers in textiles are almost endless in variety. The industry involves fiber producers—growers or manufacturers; fabric manufacturers; converters—those who finish and color fabrics; designers—those who design yarns and fabrics; marketing specialists for each of these levels. Moving into the end-use area, there are product designers—those who plan the designs for the use of fabrics in apparel, home furnishings, transportation, industry, etc.; garment manufacturers; domestic manufacturers; industrial fabric manufacturers; product manufacturers for all of these end-uses; marketing specialists for the wholesale trades; marketing organizations for the retail trade, and all others involved in the transfer of the textile fiber through all of the various stages to the final consumer. 服装

The latter fields provide stimulating career opportunities in themselves: The retailer who serves them requires salespeople, buyers, merchandise managers, and related personnel who deal with textile products. The manufacturers of these products need fabric buyers, designers, production managers, salespeople, etc. The fabric manufacturers, yarn producers, and fabric finishers need knowledgeable people. The chemical industry requires skilled personnel in a variety of areas. All of these fields utilize marketing and advertising specialists to promote textiles and textile-related products.

There are opportunities in the textile industry for artists; scientists, especially in such fields as chemistry, agriculture, home economics; business experts; advertising specialists; retailers; journalists and educators.

Textile science as it now exists has developed as a natural result of the tremendous expansion in textile knowledge and technology in this century. As the textile industry has grown, so has its science grown, evolving in part as a result of the constant need to refine definitions and classes—to sharpen the tools of the trade, in other words. Textile science, like any other discipline, has its own language, terminology, and methods of classification.

纺织工业的发展

1.1.3 Brief Development of Textile Industry

The immense scope of the textile industry makes it one of the largest in the world. The textile industry continues to be a rapidly changing one, creating new fiber and yarn modifications and manufacturing them into functional and decorative products for apparel and home furnishings as well as for industrial purposes. But it is also one of the oldest. Archeological evidence indicates that textiles of fine quality were made thousands of years before the oldest preserved accounts that refer to them.

From ancient eras until the eighteenth century, all fabrics were constructed and decorated by hand. Weaving or felting took place in the home or in small workshops. Some artist-weavers of those centuries achieved great skill, and many of their products have never been surpassed. Such beautiful examples as oriental carpets and rugs, huge tapestries portraying intricate scenes, and early Coptic textiles now hang in museums all over the world as treasured exhibits.

地毯
毯子/挂毯,织锦

But the Industrial Revolution with its factory production brought a dramatic change. In 1791, Samuel Slater brought to America the secret English methods of carding and combing of yarns; and in 1814, Francis Cabot established the first completely power-operated textile plant where all steps in the manufacture of cloth were performed. For over one hundred years, although there were great improvements in mechanical methods, the raw materials used were the natural fibers, particularly cotton, wool, linen, and silk. Late in the nineteenth century, chemists created the first man-made fiber, cellulose rayon, acetate, which assumed commercial importance early in the twentieth century. But in the post World War II era of the 1940s, the genius of the chemical industry led to a wide variety of man-made fibers that in the United States, at least, surpasses the natural fibers in importance. The spectacular changes in production have been supported by changes

粗梳/精梳
人造丝/醋酯纤维

in customer demand, growing out of increases in population and in per capita income, by changes in life-styles, and by the demand for new end-uses to which textile products might be put.

Since 1960, major changes in the production and manufacture of fibers, yarns, and fabrics have come about. Second-generation, third-generation, and fourth-generation modifications and adaptations of man-made fibers have affected behavior, use and care. New procedures in wet processing, chemical finishing, and dye chemistry provide consumers with products that meet an increasing range of old and new needs. Continued research will, it is hoped, meet the challenges of the last two decades of the twentieth century. Major changes have been occurring in the type of machinery that is used in the production of textile yarns and fabrics. 染料,染色

1.2 Textile Education 纺织教育

A distinction may be made between providing information for the customer or the consumer that the former may represent and acquiring an education. Information provides the customer with knowledge about a particular product or service offered for sale, whereas education teaches the individual how to make an intelligent choice among the many products and services available in the market. Ways in which customer education are provided are discussed in the points that follow.

1.2.1 Teaching Salespeople 售货员培训

In the small dry-goods store, salespeople still acquire much textile information through experience. This method of learning is often a slow and discouraging process. A shortcut to learning is the classroom instruction in textiles that training departments in retail stores may give their salespeople, supplemented by regular meetings conducted by the buyer of the department. 售货员 采购员

In the larger department stores and chains, the prime source of textile information comes from pamphlets, films, and fact cards provided by the manufacturers, vendors, and textile associations. More sophisticated electronic communication systems, such as closed-circuit television, which may be beamed throughout a store's branch units from a central television studio in the main or flagship store, are used in larger retail chains. The essential information is usually provided by the buyer or department manager at morning meetings. Emphasis is naturally given to new items added to the assortment. In chain stores, the pertinent information may be compiled at headquarters and distributed to the department managers in the various units. The store's training department may be responsible for writing informative booklets and providing the buyers with teaching materials for the main store and branches and for chain store units. Sometimes, the training department gives "reinforcement" instruction. When a new product appears in stock, the manufacturer's representative may present pertinent facts about it to salespeople. Part-timers are usually given the same textile information as the regular salespeople by the same executives. 小册子 卖主

Many progressive stores encourage their sponsors and junior executives to take textile courses in high schools, colleges, museums and other outside institutions.

Stores may pay part or all of their employees' tuition if they complete the courses successfully. Textile experts from outside organizations also conduct courses in retail stores.

1.2.2 Education in the College

高等院校教育

At the four-year college level, as well as in many junior and community colleges (two-year programs), the study of textiles is an integral part of the retailing curriculum. Quite often these programs are in the home economics, human ecology, and consumer science areas of the college and carry such names as fashion merchandising, fashion retailing, and consumer marketing. Although generally not found as a major program in business schools, retailing also is and has been for many decades a traditional course offering in business curricula.

The majority are located in Business Education, Distributive Education, and Distributive Education and Marketing divisions, while in some states the programs are also found in Home Economics and Industrial Arts areas.

1.2.3 Collegiate Internship Programs

学生实习

More recently, the pendulum has swung back as college administrators and faculty are coming to realize that experience in the "real world" is a valid educational experience. Moreover, some colleges grant academic credit for prior experiences if proper documentation can be provided. Students are encouraged to find part-time related employment, often being urged by faculty advisors to acquire "on the job" summer work experiences to supplement their regular college program. The student with textile background knowledge will obviously have an edge over others while competing in this expanding part-time job market.

1.3 Assisting and Guiding the Consumer

对消费者的协助与引导

1.3.1 Consumer Rights and Responsibilities

消费者的权力与义务

保养，打理

耐久压熨

免烫针织物

Today consumers want fabrics that are not only attractive, but also comfortable and durable, while requiring a minimum amount of care. These somewhat conflicting desires have been satisfied to a large extent by tremendous scientific advances in finishing technology; thus consumers find fibers and fabrics in the form of "durable press", "non-iron knits", and others that give good service, need little care, and at the same time often are beautiful in appearance.

Accordingly, to be a wise buyer in the market, consumers must acquire a great deal of knowledge that their predecessors did not need to know. The same is true of those who sell to those customers: salespeople, advertisers, and all those who market textiles. Much of this essential knowledge must come from the producers of the fibers, yarns, fabrics, and the finished product; but it must also come from the experience of knowledgeable consumers who observe closely how the products perform in relation to their intended use.

Consumers and sellers alike should understand the rights and responsibilities of the consumer in connection with the selection of textile products (see Fig. 1 – 1) and be armed with the knowledge to assure the realization of these rights.

Fig. 1 – 1 Consumer rights and responsibilities

1.3.2 Recognition of Consumers' Buying Motives 消费动机识别

For some consumers, a brand name of merchandise may be a motivating factor in selection; For other consumers, factual information is demanded for satisfaction.

If consumers can judge the wearing quality of a fabric, they are more likely to 穿着质量 get their money's worth. To determine wearing quality, one must recognize the inherent characteristics of a fabric, such as the kind of used raw material (fibers), the strength and evenness of the yarns, the construction or weave, and the perma- 强度/组织 nency of the dye or the finish. Textile education attempts to teach the consumer:

(1) To recognize and interpret the inherent characteristics of a fabric in light of its intended use.

(2) To judge the wearing quality in relation to the price.

A customer who comes to purchase a textile fabric may not ask a single ques-

tion; yet the salespeople who can determine the customer's likes and dislikes through conversation and sales talk will usually make a sale. The salesperson can also determine whether the customer is trying to satisfy physical and social needs. The customer's self-image is a consideration. Customers who believe themselves to
时尚引领者/高档
be fashion leaders may want something new, different, high style; the others will conform to what they recognize as acceptable.

有鉴别力的消费者
The discriminating consumer has at least some of the following questions in mind when buying a fabric:

适合
(1) Is this suitable to my needs and wants? (suitability)

多功能
(2) Can it be worn for a number of different occasions or purposes? (versatility)

方便
(3) Will it conserve time and effort? (convenience)

耐用
(4) Will it wear well? (durability)

舒适
(5) Will it be warm in winter? Is the texture suitable? (comfort)

保养
(6) Will this material be easy to dry-clean, launder, protect from moths, mildew, and so forth? (care)

安全
(7) Is the article safe to use, either flame-resistant or nonirritating to the skin? (safe)

相称
(8) Is the fabric good-looking? Will it look good on me? Will it go well with other garments or with the surroundings in which it will appear? (becomingness)

外观
(9) Is it in fashion? (appearance)

价格
(10) Does the price come within my means? Is it a "good buy"? (price)

情感
(11) Do I want to buy this merchandise because of associations it calls to mind? Because someone else has something like it and because I don't want to be out of style? (sentiment)

独有
(12) Will ownership of this merchandise give me a sense of possessing something unique and of great rarity? (pride of ownership)

公认
(13) Will people be impressed with my selection? Is it in line with what my group recognizes as acceptable? (recognition)

创新
(14) Does this merchandise satisfy a creative urge, particularly in buying yard goods and accessories? (creativity)

整体效果/协调关系
In question 8 the customer is considering the "ensemble idea" – the harmonious relationship of fabrics, color, and fashion-rightness, not only among various units of apparel worn together, but also among home furnishings. The customer may also ask, "Is my selection in line with my self-image?"

In answer to question 9, if the fabric is in accord with prevailing tendencies and modes of expression, if it has beauty, becomingness, and fashion-rightness, it looks good in use.

In question 10, the customer is also asking, "Am I getting my money's worth?" If the customer is limited in the amount of money that can be spent for an article, the price is of great importance as a buying motive.

饰边,花边
Many people buy old tapestries, laces, and rugs, not necessarily because of durability, but for the satisfaction of acquiring collections of fabrics. Other purchasers want materials that call to mind pleasant associations. For example, to a man whose childhood was spent in Asia Minor, where beautiful rugs are woven, rugs

would recall boyhood. Such a person's buying motive is based on sentiment.

Rivalry is an instinct, and the striving to equal or excel forms an appeal to the customer who buys to "keep up with the Joneses". This customer thrives on recognition.

Few consumers need to consider all these factors before making a purchase. To one customer, price is paramount; to another, style; to another, becomingness; to another, possibly comfort. Knowing the factors inherent in the merchandise, the salespeople can relate the factors to the buying motive.

1.4 Summary and Scope

总结与展望

The well-informed consumer is the one who can recognize and interpret the inherent characteristics of a textile fabric in the light of its intended use. With knowledge of facts about the goods, consumers can judge its probable wearing quality and can determine whether they are getting their money's worth.

Consumer education in textiles may be obtained through courses in merchandise information. The consumer is being protected and informed as a result of legislation.

Retailers have also realized that the consumer depends upon the salespeople for merchandise information, and therefore they have developed training programs toward that end.

Discriminating consumers are those who plan to get their money's worth. A person who gets value for the price paid is the one who knows the characteristics inherent in a fabric that affect the qualities of suitability, versatility, durability, convenience, comfort, care, safety, becomingness, appearance, price, sentiment, pride of ownership, recognition and creativity.

At the present time, reduction in energy consumption would appear to be vital to the future of any industry including the textile industry. Students of textiles should give thought to this increasing concern and aid in the search for solutions that will help both the industry and the consumers meet the need for reduced energy consumption and for changes in the patterns of the use of energy-rich resources in the manufacture of fibers and finishes.

能源消耗

Environment pollution such as noise pollution and pollution of natural resources such as water and air are another problems that must be faced by textile manufacturers and processors. The textile student also should at least recognize these potential dangers.

环境污染/噪声污染

Manufacturers are turning to looms that produce less noise, to the modification of equipment to reduce noise, and to protective devices for employees who must work in areas where is a high noise level.

织机

In the future, however, it is equally important that consumers and students of textile science recognize that the typical fabric available may differ from that available now in many ways, such as appearance, texture, and care procedures. It would be desirable for the student of textile science to become involved in the identification of what changes will be desirable, what changes will be demanded, and how consumers can respond to whatever changes occur. The textile specialist will be called upon to help identify patterns of change that will be of importance to the production, the selection, the use, and the care of textile products in the decades ahead.

思考题 Review Questions

1. Name five industries which are dependent upon or related to the textile industry, and indicate how they are related.
2. Name five areas of employment in the textile industry.
3. Describe your understanding to
   (1) the range and markets of textile industry.
   (2) the careers in textiles.
4. Why is the knowledge of textile fabrics important to the consumer?
5. How can the customer acquire merchandise information?
6. What is your understanding to the Consumer Rights and Responsibilities?
7. What is meant by the phrase "getting your money's worth"?
8. (1) List the questions a customer may have in mind when he or she comes to purchase a textile fabric.
   (2) Describe the most important question(s) to you when you plan to buy underwear, hosiery, a dress, a suit, a coat, handkerchiefs, or curtains.
9. Explain what is meant by sentiment or recognition as a buying motive separately.

实训题 Suggested Activities

1. List ten different articles of apparel in your home, note from their labels the kind(s) of fiber from which each are made, and indicate the fiber trademarks, if any.
2. (1) Clip an advertisement for a particular item of clothing, such as a dress, suit, hosiery, underwear, or hat.
   (2) Underline all terms descriptive of the merchandise.
   (3) What are the chief merits of the advertisement?

Chapter 2 Textile Fibers
纺织纤维

2.1 Introduction 引言

2.1.1 What are Textile Fibers 纺织纤维的定义

The term textile fiber refers to any product capable of being woven or otherwise made into fabrics. These broad reciprocal definitions are now accepted and used by nearly everyone who works with fibers or fabrics at any stage of manufacture or processing. Fibers are the smallest visible components of most fabrics.

Textile fibers may be staple or filament. Staple fibers (短纤维) are relatively short, measured in millimeters or inches. Filament fibers (长丝纤维) are relatively long, measured in meters or yards. Most natural fibers are staple; the only natural filament fiber is reeled or cultivated silk (家蚕丝). On the other hand, all manufactured (formerly termed "man-made") fibers can be staple or filament; they begin as filament, and in this form can give silky fabrics. They can also be cut or broken into staple to give fabrics that look and feel more like wool, cotton, or linen (亚麻).

2.1.2 Classification 分类

It is estimated that the number of textile fibers is more than 1,000, but this number will be increased in the future due to the development of synthetic fibers (合成纤维).

When the study of textile fibers was began, textile fibers were classified into the categories of animal, vegetable, and mineral matter, based on a systematic arrangement of fibers. With the development of man-made fibers this early classification was made obsolete, and new systems had to be devised.

In 1960, the Textile Fiber Products Identification Act (usually referred to as TFPIA) became effective. The classification system proposed in this text in Tables 2 – 1 and Table 2 – 2 indicates generic terms, in Table 2 – 2, with an asterisk and with a lowercase initial letter.

Most generic groups include trademarked fibers — those bearing a word, letter, device, or symbol that points distinctly to the origin or ownership of the particular fiber. These trademark names advertise the company as well as indicating a particular product. Trademark names listed in Table 2 – 2 are capitalized.

Table 2 - 1 Natural Fibers

Fibers	Classification
Cellulosic fibers	1. seed hairs: cotton, milkweed, kapok, cattail 2. bast fibers: flax, ramie, hemp, jute, sunn, kenaf, urena 3. leaf fibers: abaca, pineapple, agave: sisal, henequen, palm, New Zealand flax, yucca, palma istle 4. nut husk fibers: coir (coconut)
Protein fibers	1. animal-hair fibers: (1) wool (sheep) (2) specialty hair fibers: alpaca, camel, cashmere, guanaco, llama, mohair (Angora goat), vicuna (3) fur fibers: mink, muskrat, Angora rabbit 2. animal secretion: (1) silk: cultivated, dupioni, tussah (wild) (2) spider silk
Mineral fiber	asbestos (rock source)
Natural rubber	—

Table 2 - 2 Man-Made or Manufactured Fibers

Fibers	Classification
Cellulosic manufactured fibers	rayon*: (1) cuprammonium Bemberg (2) saponified cellulose rayon (temporarily discontinued) (3) viscose rayon: ① regular and high-tenacity rayon: Coloray, Enkrome, Fibro, Jetspun, Kolorbon, Sayfr ② high-wet-modulus rayon: Avril (Fiber 40 and 43), Prima, Xena, Zantrel 700 ③ hollow fibers: Viloft SI
Modified cellulosic manufactured fibers	acetate*: (1) secondary acetate: ADA (flame-resistant acetate), Celanese, Celaperm (solution dyed), Chromspun (solution dyed), Estron, Lanese, Sayfr (flame -retardant acetate) (2) triacetate*: Arnel
Protein manufactured fibers	azlon*: casein protein zein protein peanut protein soybean protein
Noncellulosic manufactured fibers	1. condensation polymer fibers: (1) nylon: ① nylon, type 66: Antron, Astroturf, Blue "C", Cadon, Cantrece, Cordura, Cumuloft, DuPont, Ultron, Vicana, Wellon ② nylon, type 6: Anso, Beaunit, Camalon, Caprolan, Crepeset, Enkaloft, Shareen, Starbrite, Zeflon, Zefran ③ nylon, type 11: Rilsan

续表

Fibers	Classification
Noncellulosic manufactured fibers	④ nylon, type 6, 10: Nylex, Quill ⑤ nylon, aromatic types: Qiana ⑥ nylon, bicomponent: Cantrece (2) aramid * : Kevlar, Nomex (3) polyester * : A-Tell, Avlin, Blue "C", Dacron, Encron, Enka, Fortrel, Kodel, Polarguard, Quintess, Setiila, Sontara, Spectrnn, Trevira, Trevira 271 – FR (flame-retardant), Vestan 16 FR, Vycron, Wistel FR, Zefran 2. addition polymer fibers: (1) anidex * : ANIM/8 (dis-continued) (2) acrylic * : Acrilan, Bi-Ioft, Courtelle, Creslan, Fina, Nandel, Orlon, Sayelle, Wintuk, Zefran (3) modacrylic * : Acrilan, Elura, Kanekalon, SEF, Teklan, Verel (4) novoloid * : Kynol (temporarily discontinued in U. S.) (5) nytril * (6) olefin * : ① polyethylene ② polypropylene: Chevron, Eastman, Fibrilon, Herculon, Marvess, Meraklon, Polymacrame, Typar, Vectra, Poly-Ti (7) saran * : Saran (8) vinal * : Cerniona, Kuralon (9) vinyon * : Rhovyl, Valren, Vinyon HH, Voplex 3. elastomers: (1) spandex * : Cleerspun, Glospan, Lycra (2) rubber * : Contro, Lastex (3) lastrile *
Mineral manufactured fibers	1. glass * : Beta, Chemglass, Fiberglas, Modiglass, PPG 2. metallic * : Alistran, Brunsmet, Chromflex, Chnimel-R, Lurex, Malora, Metlon, X-static
Other manufactured fibers	1. alginate 2. inorganic: Avceram — carbon silica, Boron, Fiberfax — alumina silica, quartz fibers, sapphire whiskers, Thornel — graphite 3. organic fibers with no TFPIA classification: PBI polybenzimidazole, Solvex — polycarbonate, Cordelan — polychal (biconstituent), Teflon — tetrafluoroethylene, Urylon— polyurea 4. biconstituent: Arnel Blend 100 — triacetate and polyester, Arnel Plus — nylon and triacetate, Mirafi 140 — nylon and olefin, Mon-velle — nylon and spandex, Raycelon — rayon and nylon, Source — nylon and poly ester, Tricelon — nylon and acetate

* Generic.

A primary listing of generics and their relationships is shown graphically as a textile fiber "family tree" in Fig. 2 – 1.

2.1.3 Fiber Properties

纤维性能

The geometrical, physical, and chemical properties of the textile fibers are very

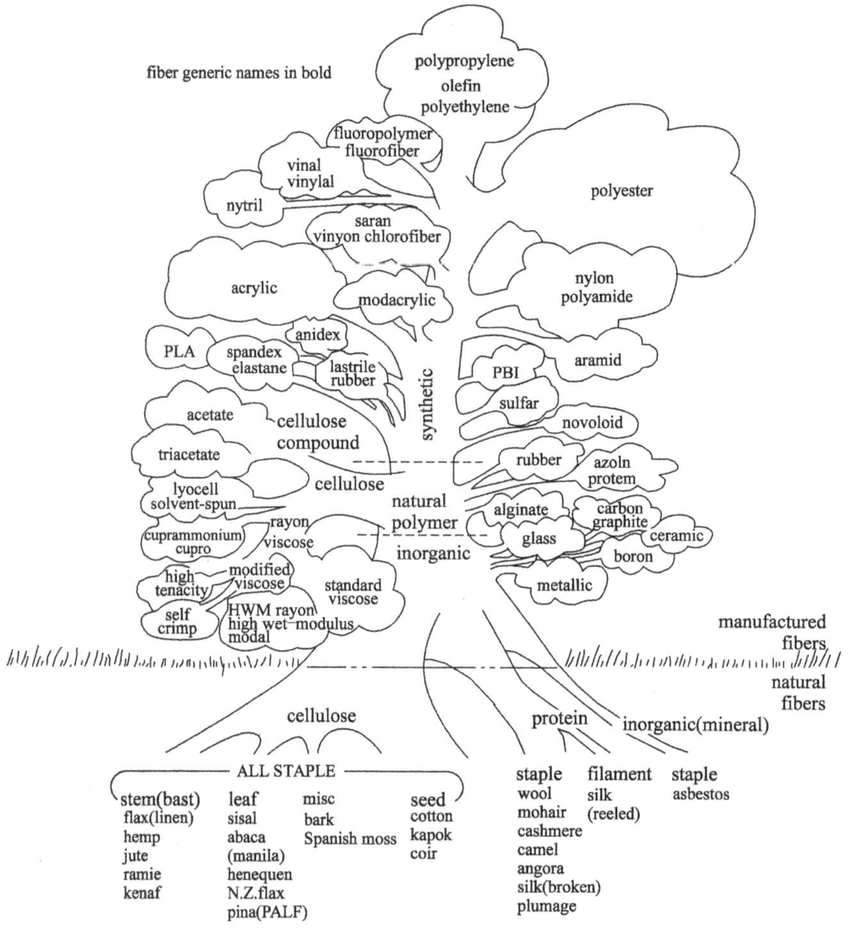

Fig. 2 - 1 Textile fiber "family tree"

important factors for processing those fibers into any kind of textile article. These primary properties include a high length-to-breadth (width) ratio, tenacity or adequate strength, flexibility or pliability, cohesiveness or spinning quality, and uniformity.

Secondary properties are those that are desirable; they may, and frequently do, improve the consumer satisfaction with the textile made from the fiber. Characteristics in this group include physical shape, specific gravity, luster, moisture regain, elastic recovery, elongation, resilience, thermal behavior, resistance to biological organisms, and resistance to chemical and other environmental conditions.

Additional properties that affect fiber performance include the way the fiber reacts to selected chemicals, to environmental conditions such as sunlight and other climatic variables, to microorganisms such as bacteria and fungi, and to insects such as moths and carpet beetles.

2.2 Natural Fibers

天然纤维

2.2.1 Cellulosic Natural Fibers

天然纤维素纤维

Plant fibers are composed largely of cellulose, and therefore are classified as cellulosic or vegetable natural fibers. The term natural cellulosic is preferred, for it indicates the simple chemistry of the substances and provides a scientific method for comparing natural cellulosic with man-made cellulosic fibers.

纤维素/天然纤维素或植物纤维

Cellulose is a linear polymer or long-chain molecule built by combining several thousand anhydroglucose units. Although glucose, a simple sugar, is soluble in water, cellulose is not because of the immense size of the molecule. Cellulose is a carbohydrate; it contains the elements carbon (44.4 percent), hydrogen (6.2 percent), and oxygen (49.4 percent). The repeating units can be diagramed as shown in Fig. 2-2.

线型聚合物/长链分子/葡萄糖酐单体

碳水化合物

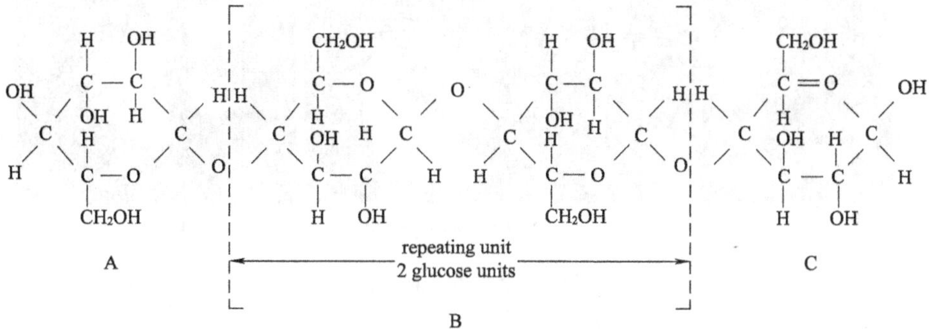

Fig. 2-2 Diagram of a cellulose molecule

Two glucose units combine first to form cellobiose; then many cellobiose units combine to form cellulose. The number of anhydroglucose units in the cellulose molecules is referred to as the degree of polymerization of the molecule. Each repeating unit equals to 2 anhydroglucose units. The glucose units are flipped over as they join together.

纤维二糖

聚合度

Cellulose fibers have several properties in common. They burn easily and quickly with a yellow flame; they give off a smell like that of burning paper or leaves; they deposit a light, fluffy, grayish residue or ash. Cellulose is damaged by acid solutions, especially strong mineral acids, but it possesses excellent resistance to alkaline solutions.

蓬松的/浅灰色残渣

耐碱性

In general, cellulose is low in elasticity and resilience; consequently, it wrinkles excessively. Cellulose fibers are soft and absorbent, so they usually make comfortable products. The fibers launder readily and can withstand strong detergents, high temperatures, and bleaches (if used properly). This group of fibers is seldom damaged by insects, but fungi, such as mildew, will destroy cellulose or stain it severely.

棉 2.2.1.1 Cotton

Cotton is the fiber used most in the world. However, during the last decades world consumption of cotton has dropped.

生长、生产和加工 (1) Growth, Production and Processing

锦葵属植物 Cotton is in the Mallow family (Malvaceae), genus Gossypium, and there are many species. The best types for textiles develop fibers called mature cotton or
棉绒 lint, which can be spun into yarn; other types produce only immature cotton or
棉短绒 linters, also called fuzz, that cannot be spun into yarn.

Cotton seed is planted, weeded, and cultivated during an 8 – 12 weeks period until the plant, grown to a height of 1 – 2 m, forms flowers. These are cream colored at first, then, over 3 days, turn to deep red and the petals fall, leaving the
棉桃 seed pod or boll, which is about the size of an egg. A boll contains about 30 seeds, and fibers (each a single cell) grow out on the surface of each seed (See Fig. 2 – 3).

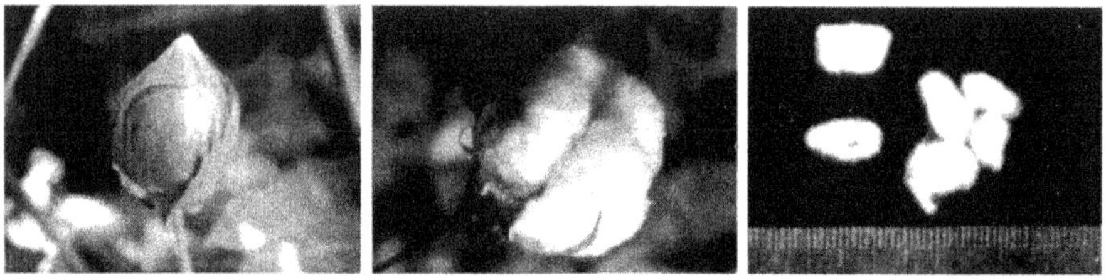

(a) Cotton boll (b) Open boll with fibers (c) Cotton seeds

Fig. 2 – 3 Growth of cotton

During the first 24 days after pollination, a fiber develops only in length, folding back on itself inside the closed boll and reaching a length up to 1,000 times its di-
初生胞壁 ameter. During this time, the fiber has only a thin primary wall, lying under the
表皮 very thin, outer, waxy cuticle. Fibers that start to grow in the first 2 – 3 days after
次生胞壁 flowering will go on to develop the secondary wall of mature cotton fibers (lint); those that start to grow later never develop more than the primary wall and are called immature cotton, linters, or fuzz. They cannot be spun into yarn for textiles, but are used for cotton batting or as a cellulose base for manufacture of rayon.

After 24 days, then, a fiber does not grow longer but begins to thicken, maintai-
中腔 ning a central canal (lumen) carrying the plant juices that nourish it. Each day, a layer of cellulose is laid down in the fiber in a spiral arrangement that reverses each day; this thickening of the secondary wall continues for about 24 days more, when the boll opens. The fibers, which have been almost cylindrical in shape, dry, flat-
反向转曲 ten, and twist to form the reversing convolutions so typical of cotton seen under the microscope. These folds and twists are very important in spinning fine yarns, and are one of the main reasons why such a (relatively) short fiber can be spun into yarn—the only seed fiber that provides a textile fiber among so many thousands of types.

Cotton in the open boll is almost pure cellulose available for use. While some waxes and pectins, protein, coloring matter, and minerals are present, only 4 – 12 percent of the fiber is anything other than cellulose. It needs only to be picked (mostly by machine today) and separated from the seed, a process carried out in the cotton gin.

蜡质/胶质
摘棉
轧棉机

Cotton can be picked by hand, by mechanical picking machines, or by stripping devices. Hand picking results in more uniform and better-quality cotton, because the pickers can select the mature fibers only and can repick the fields as many times as it is profitable.

After the cotton is picked, it is taken to the ginnery, where the fiber, called cotton lint by the trade, is separated from the seed. The cotton gin pulls the fibers from the seed and carries them away, but there is also quite a lot of other plant matter picked up by mechanical harvesters which has to be eliminated at some stage. Some gins do this by passing cotton through a huller front, but in newer gin stands, the seed cotton drops directly into a roll box follow the arrow in Fig. 2 – 4 (a), where the seed is separated from the mature cotton fiber (lint). When all the long fibers are removed, the seeds slide down the face of the ginning rib between the saws, and are taken away on a conveyor. Revolving circular saws pull the lint through closely spaced ribs that do not allow the seed through. Lint is removed from the saw by a rotating brush (or a blast or air) and goes on to Fig. 2 – 4 (b), the lint cleaner. Here immature seeds (motes) and bits of plant matter (trash) are removed. The lint cleaner also blends and aligns the fibers. Cotton seeds are pressed to give an important polyunsaturated oil, while their hulls are used for animal feed or fertilizer. Some trash gets in with the usable lint fibers, and can be seen in a fabric such as unbleached muslin. The fibers or cotton lint

轧棉厂
多不饱和油/外壳
薄纱织物，平纹细布

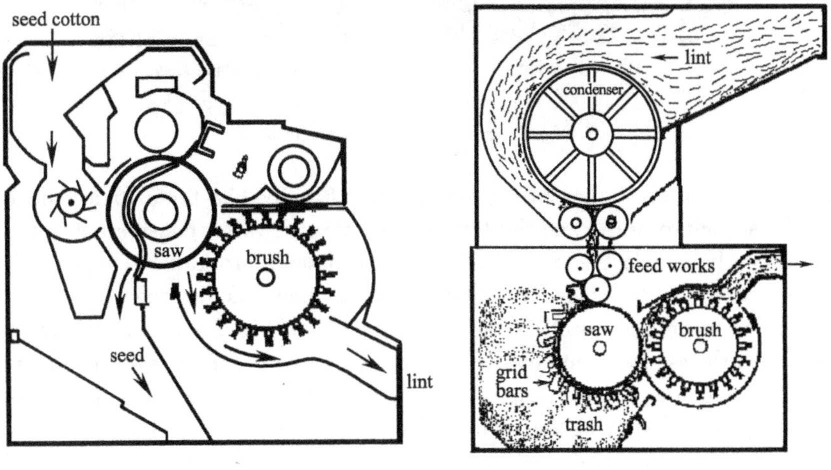

(a) A modern, high capacity gin stand without a huller front

(b) A typical saw-type lint cleaner which removes trash, and blends and aligns the fibers

Fig. 2 – 4 A gin stand and a lint cleaner

are packed into large bales at the ginnery. Each bale weighs about 227 kg (500 pounds) gross.

成熟度
开棉机

The fineness and maturity of cotton fiber are important criteria in determining quality and contribute to yarn strength and appearance of yarns and fabrics. This measure is determined by several instruments all of which use the same scale and report the measurement as "micronaire reading". To maintain an average range of fineness, mills blend various bales of fiber to provide a consistent average fineness for input to the picker.

一束纤维
单纤维

Cotton is rated or evaluated on the basis of several additional properties. These include: fiber strength, uniformity, cohesiveness, pliability, elastic recovery, and resiliency. These properties are usually determined with a bundle of fibers rather than individual fibers.

形态结构

(2) Morphology

Cotton fibers are composed of an outer cuticle (skin) and primary wall, a secondary wall, and a central core or lumen. Immature fibers exhibit thin wall structures and a large lumen, whereas mature fibers have thick walls and a small lumen that may not be continuous, because the wall closes the lumen in some sections [Fig. 2 - 5 (a)].

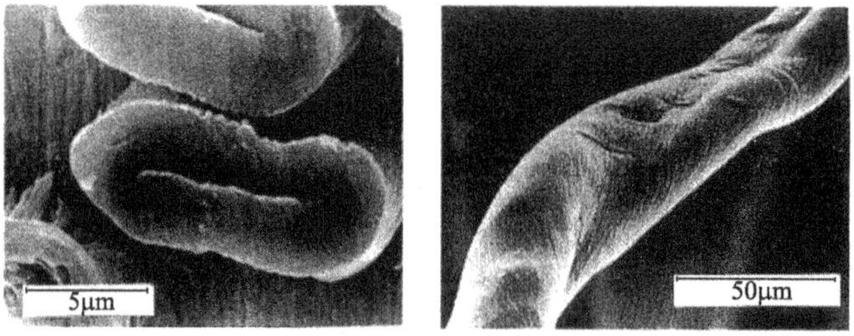

(a) Cross section　　(b) Longitudinal view

Fig. 2 - 5 SEM of regular cotton

腰圆形

The cross section of the fiber (Fig. 2 - 6) usually has three areas: the outer skin, the secondary wall, and the lumen. The contour varies considerably: some fibers are nearly circular, some are elliptical, and some are kidney-shaped. Immature fibers are generally more irregular in contour than mature fibers.

丝光处理

Fibers that have been swollen, as in mercerization (Fig. 2 - 7), do not show the twist as clearly as does untreated fiber. Swollen fibers appear smooth and round when compared with the untreated. Immature fibers also have few convolutions.

带状
条痕

The longitudinal view of the regular fiber [Fig. 2 - 5 (b)] shows a ribbonlike shape with twist (convolutions) at irregular intervals. The diameter of the fiber narrows at the tip. The lumen may appear as a shaded area or as striations; this is more obvious in immature fibers.

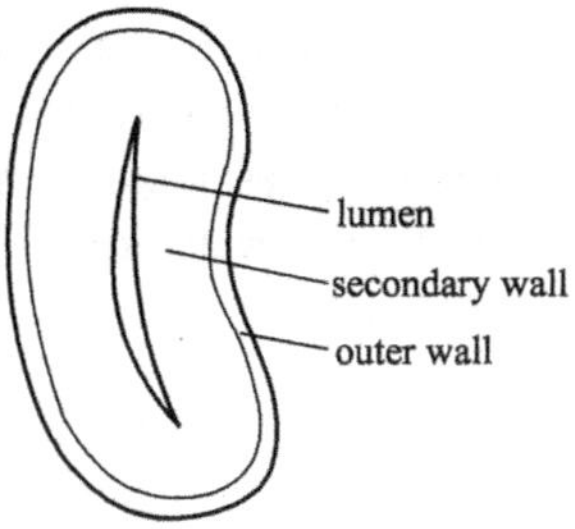

Fig. 2 – 6 Cross section of a cotton fiber

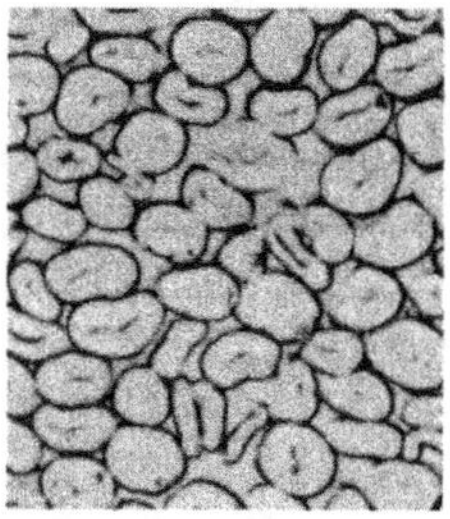

Fig. 2 – 7 Cross section of mercerized cotton

(3) Properties 性能

Cotton properties are summarized in Table 2 – 3.

Table 2 – 3 Properties of Cotton Fibers

Property		Evaluation
shape		fairly uniform in width, 12 – 20μm; length variable from 1.27 – 6.35 cm; most fibers 2.22 – 3.18 cm in length
luster		low
strength		26.46 – 44.1cN/tex (3.0 – 5.0 gf/d), medium strength
elastic		low elasticity, 75% recovery at 2% extension;3% – 10%
recovery/elongation		elongation
resiliency		low
density/ specific gravity		1.54 – 1.56g/cm^3
moisture absorption		8.5% at standard conditions. 15% – 25% at saturation, 8.5% – 10.3% absorption for mercerized
dimensional stability		considered relatively stable
resistant to	acids	damage, weaken fiber
	alkalis	resistant, no harmful effects
	organic solvents	high resistance to most
	sunlight	prolonged exposure weakens fibers
	microorganisms	mildew (fungi) damages fibers, rot-bacteria damages the fibers
	insects	silverfish damage cotton
thermal reactions	to heat	decompose following extended subjection to heat at 150℃
	to flame	burn readily

As well as resisting alkaline agents, cotton, which has good strength when dry, gets about 25 percent stronger when it is wet. However, cotton is certainly not easy care without special durable-press treatment, since it wrinkles readily, swells in water so that fabrics made of it shrink, and can mildew, as anyone knows who has left a damp towel lying about. Cotton fabrics also lose a crisp press when damp (poor wet press retention).

耐久压熨整理

2.2.1.2 Flax (Linen)

亚麻

Flax or linen is one of the oldest and most-prized natural fibers. In textiles study, the fiber is termed flax, while the fabric made of it is called linen. The fiber bundles lie between the outer bark and the core of the stem of Linum usitatissimum—the meaning of the Latin is "the most useful linum"; many of the Linum genus grow wild.

亚麻布
亚麻属植物

(1) Growth, Production and Processing

生长、生产和加工

Flax is a bast fiber—a woody fiber obtained from the phloem of plants. It derives from the stalk or stem of the Linum usitatissimum.

韧皮部
茎,梗/茎,干

The flax plant requires a temperate climate with generally cloudy skies and adequate moisture. Bright sunlight and high temperatures are damaging unless alternated with abundant rainfall. The plant must obtain considerable nutritive value from the soil, and thus, best-quality flax requires a crop-rotation program of 5 years. Another variety of flax is grown for oil from its seeds; linseed oil has long been used as a quick-drying oil base for paints.

亚麻子油

Flax for fiber is pulled by hand or by mechanical pullers to keep the roots intact, for the fibers extend below ground surface. Harvesting occurs in late August when the plant is a rich brown color. After drying, the plant is rippled; that is, it is pulled through special threshing machines that remove the seed bolls or pods.

Next comes the central process, retting, which is really rotting—a natural breakdown of the stem materials and pectins that bind the fiber bundles to the stem. Dew, pool, or river retting takes 4 – 6 weeks. Lukewarm water retting in tanks takes 3 – 5 days, after which the bundles are dried for 1 – 2 weeks. Chemical retting speeds up the process, but when the stem materials are so attacked, the fibers are also weakened and harshened. Top quality fibers today are often still dew retted, where the cut flax is left in the fields to let the action of moisture from dew and the action of microorganisms break down the outer stem.

沤麻

After retting and drying, the stems are broken up by scutching with wooden breaker rollers, then the straw is beaten out as waste or shins for use in chipboard. The fibers, separated into "hands", are then hackled or combed to take out short waste fibers called tow, and to lay parallel the long fibers called line; these may reach 500 mm long (See Fig. 2 – 8).

打麻
梳理
短麻/长麻

(2) Morphology

形态结构

Flax fiber is composed of fibrils or bundles of fiber cells held together by a bonding or gummy substance (Fig. 2 – 9). Under the microscope the longitudinal view of the fiber shows the width to be quite irregular [Fig. 2 – 10 (a)]. The central canal or lumen casts a shadow, giving a slightly darker effect down the center.

微纤维,原纤

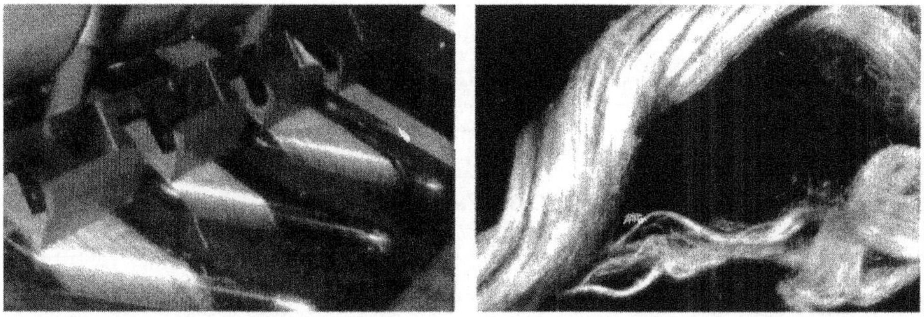

(a) Hackled(combed)flax ready for spinning　　(b) Line flax,resembling glossy human hair

Fig. 2 –8 Hackled flax

There are no convolutions as in cotton, but longitudinal lines or striations can be seen. The points at which the fiber width changes are marked by swellings and irregular joint formations called nodes. These are similar to the joints on bamboo. 竹节

The cross-sectional view [Fig. 2 – 10 (b)] clearly shows the lumen, the thick outer wall, and a somewhat polygonal shape. Immature flax may be oval in shape and usually has a larger lumen than mature fiber.

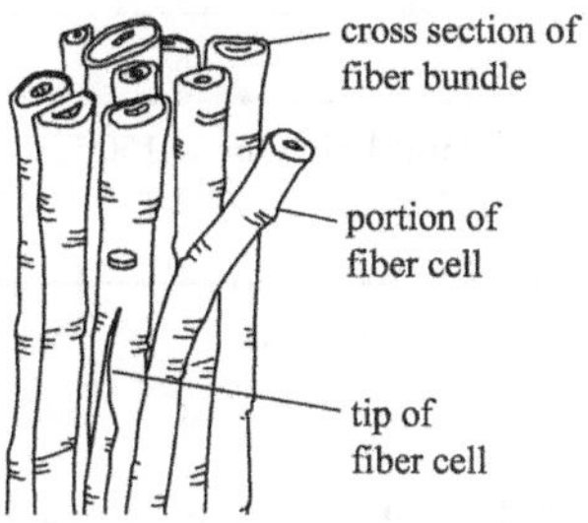

Fig. 2 –9 Schematic diagram of fibers in flax fiber

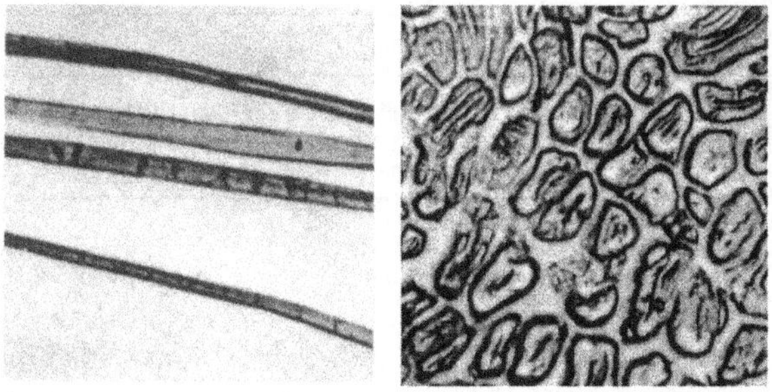

(a) Longitudinal view　　(b) Cross-sectional view

Fig. 2 –10 Photomicrograph of flax

(3) Properties 性能

Flax properties are outstanding — it is a fiber that is still valued after some 10,000years. It has better dry strength than cotton, and like cotton it gets 25 percent stronger when wet. It absorbs more moisture (for comfort and static resistance), and it wicks; it is longer, smoother, and more lustrous than cotton.

Flax properties are summarized in Table 2 –4.

Table 2 – 4 Properties of Flax Fibers

Property		Evaluation
shape		width varies, mean diameter is about 21.17μm; length of fiber varies from a few cm to 55.88 cm or more; average length after processing, 25.4 – 38.1 cm
luster		high
strength		average is 48.51 – 57.33 cN/tex (5.5 – 6.5 gf/d); range is 22.93 – 67.91 cN/tex (2.6 – 7.7 gf/d)
elastic recovery/ elongation		elasticity recovery 65% at 2% extension; low elongation 2.7% – 3.3%
resiliency		poor
density		1.50 g /cm^3
moisture absorption		good, approximately 12% at standard conditions
dimensional stability		good
resistant to	acids	good to cool, dilute acids; low or poor to hot, dilute; poor to concentrated either hot or cool
	alkalis	high resistant
	sunlight	good
	microorganisms	mildew will grow on and damages fibers, particularly in humid atmosphere
	insects	good
thermal reactions	to heat	gradual decomposition after prolonged exposure at 150℃
	to flame	burn readily

苎麻 2.2.1.3 Ramie

Ramie (China grass, rhea) is a minor fiber but can rival linen in strength, beauty, and absorbency. It is stiffer than flax, and is even more difficult than flax to loosen from the plant stems without chemical treatment, which harshens the fiber. The ramie called China grass is grown in China, while the type called rhea comes from Malaysia, Africa, and Mexico. A good deal of ramie appeared on the North American market in the 1980s, blended, usually with cotton, in both knit and woven fabrics.

After cutting, the ramie stalks are decorticated by one of several machine or hand methods to remove the outer woody covering and reveal the fine fibers. These
脱胶 are degummed to eliminate the pectins and waxes. Degumming is usually done by
苛性钠/漂白/中和 soaking the fibers in caustic soda for several hours, then bleaching and neutralizing in a dilute acid bath. Finally, the fibers are washed and dried.

Ramie fibers are long and very fine. They are white and lustrous and almost silklike in appearance. The strength of ramie is excellent and varies from 45 – 56 cN/tex 5.3 – 7.4g per denier. Elastic recovery is low, and elongation is poor. The fibers are somewhat stiff.

2.2.1.4 Hemp 大麻

Hemp is an old bast fiber of the Mulberry family Moraceae, and like flax, the fiber is removed from the plant stems by retting. Hemp has many of the good properties of flax, and is one of the strongest natural fibers, even when wet. It has been used mostly for coarser products, such as strong twine and other cordage, fish 麻线/绳索 nets, canvas, and a better quality of burlap. It can give a fabric hard to tell from 帆布/粗麻布 linen, and historically, fine fabrics as well as good paper have long been made from hemp in China. Hemp seed oil compares in yield to rape or sunflower oils, and the woody core of the stem can be made into building materials as well as paper, so there are nontextile uses as well. Plants grow quickly in temperate climates to heights of some 4 m in 3 months, resist weeds, and need little application of pesticides or agrochemicals. The large root system aerates the soil it grows in. However, hemp has more lignin than flax, and the fiber bundles are harder to sep- 木质素 arate to give fine yarns.

2.2.1.5 Jute 黄麻

Jute is one of the most easily spun and cheapest fibers but is also one of the weakest, most brittle, and least durable. It comes from plants of the Corchorus genus, grown mainly in India and Bangladesh, but also in China, Nepal, and Thailand. Fibers range in length from shorter than 25 mm to about 300 mm. Retting takes 2 – 3 weeks, and fiber yield is less than 10 percent of the total stem weight. We have used a great deal of jute for coarse twine and cordage and as burlap in sacking 麻袋布 and carpet and linoleum backing. Jute cannot be used in water (e.g., for fish 油布衬 nets) as it deteriorates (rots) quickly when wet.

The International Jute Organization is promoting jute bags for the food industry, produced using a vegetable jute batching oil instead of a petroleum-based oil. Fur- 翻砂用油,造绳缆用油 thermore, as the use of jute in sacking, cordage, and especially carpet backing has been replaced extensively by polypropylene, other uses for it have been active- 聚丙烯 ly sought, especially in geotextiles, such as fabrics used to prevent erosion on hill- 土工布 sides.

2.2.1.6 Kapok, Coir, Kenaf, ect. 木棉、椰壳纤维、洋麻等

Kapok, from the seed pod of a tropical tree (Ceiba pentandra), is a soft fiber that is buoyant because it is hollow; it is used for stuffing in furnishings and life 填塞料 jackets, largely replaced now by foam or polyester fiberfill.

Coir, from the husk of the coconut (Cocus nucifera), is coarse, stiff, and strong even when wet; it is used for rough matting and cordage (twine, rope), and 垫,席 in geotextiles, as it is biodegradable. 可生物降解的

Kenaf and roselle are minor bast fibers used for the same purposes as jute. Kenaf 洛神葵 is grown in India, Bangladesh, China, Thailand, and other parts of Southeast Asia.

Sisal, family Amarylklidaceae or Agavaceae, is one of the best known of the 剑麻 Agave genus of plants. The bundles of fibers, which are coarse and woody, may be 龙舌兰属植物

大手提袋 up to 1 m in length, held together by resin. Sisal is used for tote-bags, summer handbags, matting carpets, cordage, and bristles. The plant is a native of Mexico, but 80 percent is now grown in East African countries, as well as Brazil, Mexico, and Central America. These fibers cannot be used for fish nets, as saltwater degrades them.

龙舌兰叶纤维/马尼拉麻 Henequen is another of the Agavaceae family, used mostly for twine. Abaca, Manila "hemp", "sisal hemp", "sunn hemp", or Manila is a fiber from the leaves of a member of the Banana family, used for cordage, matting, and some clothing.

莲茎 Fibers from lotus stems are again being made into fabric in Burma by villagers, encouraged by their cooperative association. This is a rare cloth, with some 长袍 220,000 stems needed to make fabric for a set of monk's robes.

天然蛋白质纤维 2.2.2 Protein Natural Fibers

Protein is a major building material of animals; flesh, skin, hair, antlers, plumage, and spider webs are all proteins, differing from one another according to their function and the creature that made them.

多肽 Protein are polypeptides, and have the most complex makeup of any fiber poly- 氨基酸/肽键 mer. The units are amino acids, joined together with amide or peptide linkages. Long chains are formed, some folded, others lying in spirals or helixes. The pro- 蚕丝蛋白/角蛋白 tein of silk (fibroin) is folded; that of wool (keratin) is a single spiral or helix, 交联/胶原质 with some cross-links; while the collagen of leather is a triple helix.

Each type of protein has its own assortment of amino acids made up of carbon (C), hydrogen (H), oxygen (O), and nitrogen (N). plus, in some units of hair fibers, sulfur (S); these last two (N and S) account for the different burning smell of wool from cellulose fibers, which are made up only of C, H, and O.

Fibers in this group have excellent moisture absorbency. Their standard moisture regain is high, and they absorb additional moisture at the saturation point. Protein fibers tend to be warmer than natural cellulose fibers. The low degree of electrical conductance contributes to a buildup of static charge, so the resulting yarns and fabrics tend to release much static electricity, although this problem is reduced in the presence of moisture.

The density of fibers in this group is less than that of cellulosic fibers. As a group they have good resiliency and elasticity. All protein fibers except silk are comparatively weak, and even weaker when wet, so care must be exercised in cleaning. The fibers burn slowly in flame, are considered self-extinguishing, leave a brittle, beadlike residue, and smell like burning hair or meat.

Protein fibers have a dry, warm hand, are weaker wet. and are sensitive to alkali and chlorine bleach, but not to acid.

蚕丝 2.2.2.1 Silk

家蚕丝 Cultivated or reeled silk is the only filament natural fiber, discovered and developed in China and often called "the queen of fibers". We use other silk fibers, including silk in staple form, but when we talk about "silky", we imply continuous filament and a fine, lustrous, white fiber.

Gum makes raw silk dull and stiff and of an ecru color. It does not become silky until the gum is removed. To get such a long fiber, nearly the entire length of the very fine filament making up the cocoon of the silkworm has to be reeled off unbroken, and when a sticky gum is removed, what we recognize as silky silk emerges. 生丝 蚕茧/蚕

(1) Morphology 形态结构

Cultivated degummed silk, viewed longitudinally under a microscope, resembles a smooth, transparent rod [Fig. 2 – 11(a)]. Silk in the gum has rough, irregular surfaces. Wild silk tends to be quite uneven and is somewhat dark. It may have longitudinal striations. 熟丝,已脱胶的丝

Cross-sectional views of silk show triangular fibers with no markings [Fig. 2 – 11 (b)]. Two filaments usually lie with their flat sides together; this can be explained by the fact that two filaments are extruded by each silkworm; they come together as indicated in Fig. 2 – 12. The two are called brins. 单丝

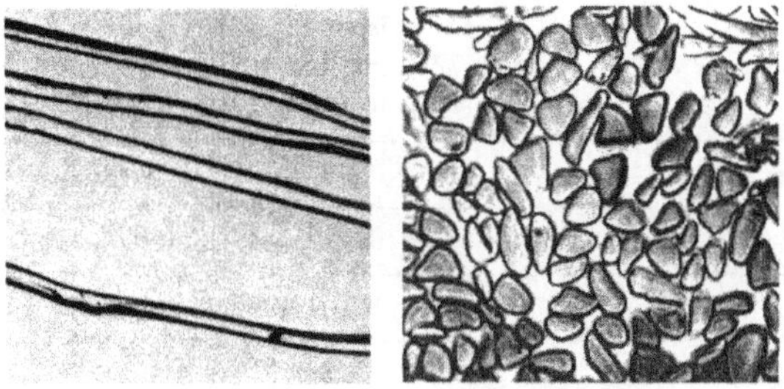

(a) Longitudinal view (b) Cross-sectional view

Fig. 2 – 11 Photomicrograph of silk

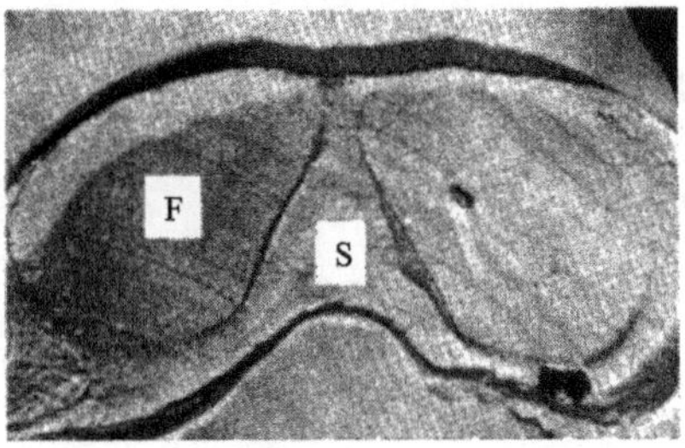

Fig. 2 – 12 SEM of silk filaments or brins
F: Fibroin, S: Sericin

(2) Properties 性能

Silk properties are unique, even though silk is a protein fiber like wool. The silk protein, fibroin, is simpler than wool keratin, without the coiling and cross-links, so silk is not as resilient as wool. It has low wet resiliency and wrinkles easily when wet — watch for this when caring for washable silk. Silk has a degree of elastic recovery. 卷绕 有回弹力的

绝缘体

It has a warm, dry hand and can be a good insulator, although we seldom use it as a fabric to keep us warm. Silk absorbs well, and is strong when dry although it does lose strength when wet.

排汗 弱酸性 氨基化合物/碱

Silk can be weakened by long exposure to sunlight. More seriously, perspiration can attack silk clothing. Proteins are not sensitive to acid but are to alkali; fresh perspiration is mildly acidic, but is acted on by bacteria on the skin, breaking down ammonium compounds to form alkali. However, it is only after time that this can concentrate and do damage to the fibers and to dyestuffs, so silks should be cleaned soon after wearing.

Table 2 –5 shows the properties of silk fibers.

Table 2 –5 Properties of Silk Fibers

Property		Evaluation
shape		The fiber is fine, 9 – 11μm in diameter; it is long from 914.4 – 1188.72m, the filaments is smooth, even white to cream color. Wild silk is uneven and dark in color
luster		high
strength	dry	21.17 –449.82 cN/tex (2.4 –51 gf/d)
	wet	17.64 –37.93 cN/tex (2.0 –4.3 gf/d)
elastic recovery/ elongation		92% recovery at 2% extension
elongation	dry	10% –25%
	wet	33% –35%
resiliency		medium
density		1.25 –1.34g/cm^3
moisture absorption 20℃, 65% R.H.		11.0%
saturation		25% –35%
dimensional stability		good
resistant to	acids	Low, dissolved or damaged by most mineral acids; organic acids do not damage
	alkalis	Strong alkalis damage fiber; weak have little or no effect
	sunlight	Prolonged exposure causes fiber breakdown.
	microorganisms	good resistant
	insects	destroyed by carpet beetles
thermal reactions	to heat	Temperatures above 150℃ result in yellowing and general discoloration
	to flame	Burns with a sputtering flame

2.2.2.2 Wool

羊毛

Wool is taken in this book to be the hair of the domestic sheep; what we will call specialty hair fibers, such as camel hair, are not usually termed wool in textiles or fashion. Domestic sheep raised for wool have been bred with less and less guardhair. the coarse outer coat that protects wild animals and is found on all fur-bearers, even domestic cats. This layer is distinct from the fine, insulating undercoat called "down". Wool from most breeds of sheep contains a portion of guardhair as coarse white fibers called kemp that do not take dye; they may be used as effect fibers in fabric such as tweed.

毛发纤维
长毛皮动物
绒毛,茸花
粗花呢

Fig. 2 – 13 shows the international Pure Wool mark and Wool Blend symbol. The Pure Wool mark is used for products containing 100 percent wool; the Wool Blend mark is a symbol used for products containing at least 60 percent wool.

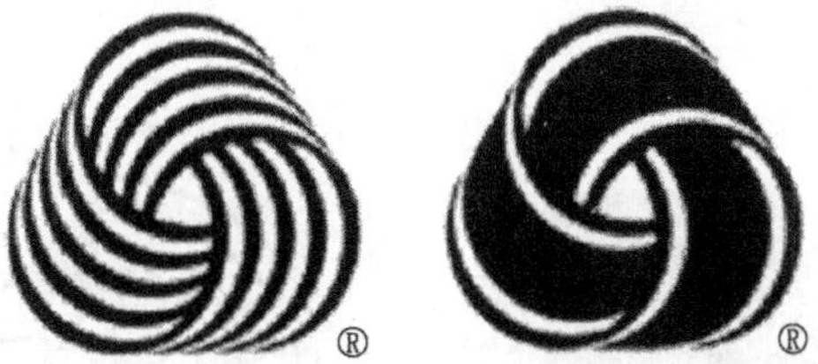

(a) Pure Wool (b) Wool Blend

Fig. 2 – 13 Logos of the wool

(1) Growth, Production and Processing

生长、生产和加工

Production of wool involves many breeds of sheep all over the world and requires a great deal of grazing land.

Sheep are sheared once a year in most countries, and the wool just shorn from a living sheep is called a fleece; Wool from the skin of a dead sheep is called pulled wool.

套毛/皮板毛

Wool is only a portion of each fleece, with up to half the weight made up of wool grease (more accurately, wax, and known as lanolin in purified form), suint (perspiration), and all kinds of dirt and plant matter, especially with the Merino (see Fig. 2 – 14 and Fig. 2 – 15). Wool is generally washed (scoured) and treated to remove plant matter (carbonized).

羊毛脂/羊毛脂/羊毛汗/美利奴细毛羊
炭化

(2) Morphology

形态结构

Under microscopic observation, the length of the wool fiber clearly shows a scalelike structure (Fig. 2 – 16). A cross section of wool shows three distinct parts to the fiber (Fig. 2 – 17). The outside layer, called the epidermis or cuticle, is composed of the scales. These scales are somewhat horny and irregular in shape, and they overlap, with the tip pointing toward the tip of the fiber; they are similar to fish scales. The edges of the scales project, so that the fiber is rougher from tip to root than from root to tip. When fibers are moved about, there is a kind of ratchet effect, so they move in the direction of their root ends; the fibers tangle to-

鳞状结构
表皮
鳞片/角状的
方向性摩擦效应

缩绒,毡缩 gether, a process called felting.

Fig. 2 – 14 Group of Merino sheep

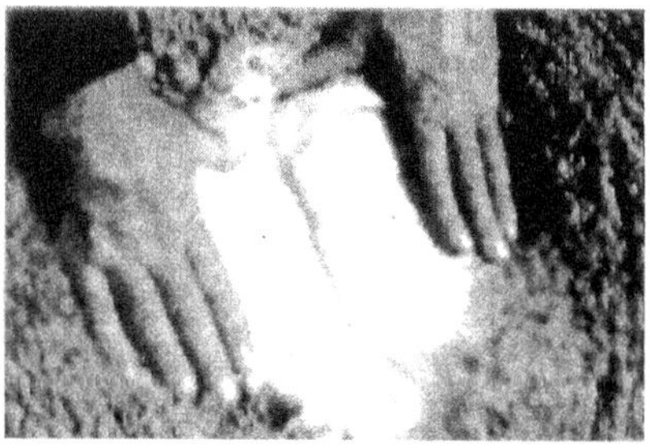

Fig. 2 – 15 Coat of a Merino sheep parted showing close, crimped fibers

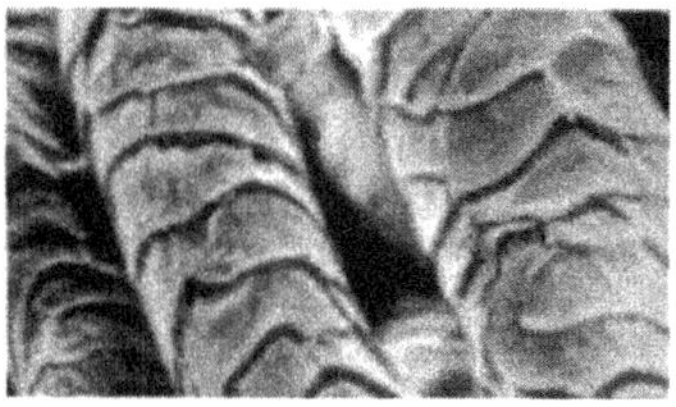

Fig. 2 – 16 Wool fibers with outer scale covering

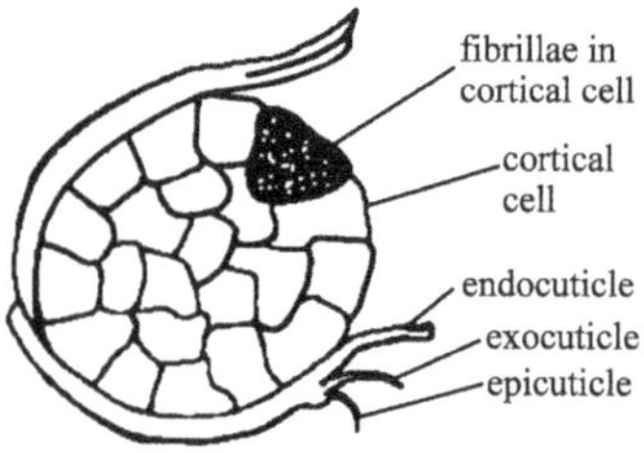

Fig. 2 – 17 Wool fiber structure: cuticle with scales, cortical cells

皮质层/皮层细胞 纺锤形 正皮质层/偏皮质层 双组分的 螺旋形/卷曲 微纤维

Under the cuticle is the cortex, the major portion of the fiber, composed of cortical cells, which can take up and hold moisture. Cortical cells are long and spindle-shaped and provide fiber strength and elasticity. The cortex accounts for approximately 90 percent of the fiber mass. Wool is the most absorbent fiber of all, and is also hygroscopic—takes up moisture without feeling wet. The cells are of two types, ortho-cortex and para-cortex, twisted together along the length of the fiber, making it bicomponent (Fig. 2 – 18); the two sides behave differently as a wet fiber dries, one shrinking more than the other, resulting in a spiral crimp (Fig. 2 – 19). The cortical cells are made up of microfibrils. which are finally composed of the coiled molecular chains.

髓质层

In the center of some coarser wools is a medulla with air spaces. The size of the medulla varies and in fine fibers may be nearly invisible. This is the area through which food reached the fiber during growth, and it contains pigment that gives color to fibers.

性能

(3) Properties

The properties of wool (shown in Table 2 – 6) result from its structure, adding up to a fiber that makes up into fabrics well suited to keep us warm in cold weather and comfortable under many other conditions.

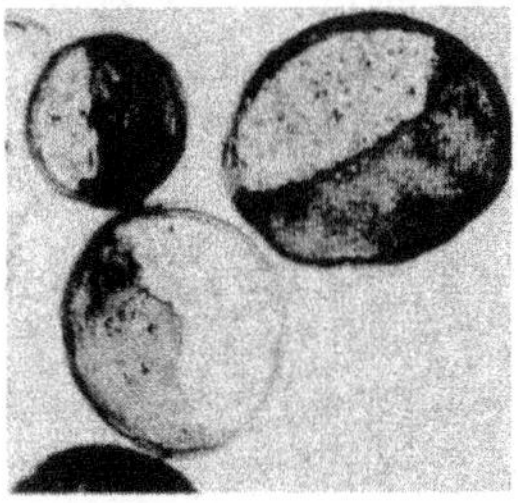

Fig. 2 – 18 Cortex of wool stained toshow the bicomponent nature of the cells

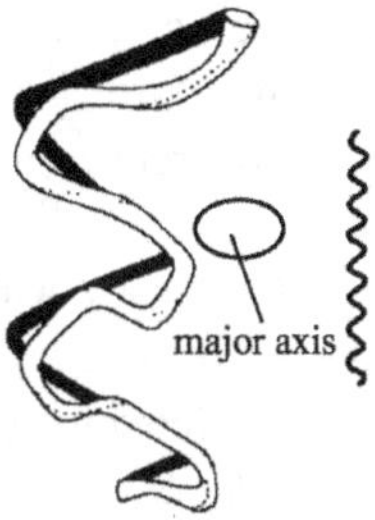

Fig. 2 – 19 Three-dimensional wool crimp

Table 2 – 6 Properties of Wool Fibers

Property		Evaluation
shape		fiber length varies from 3.81 to 38.1 cm, width varies from 15 to 70μm; fibers have a natural crimp or waviness
luster		medium to medium high
strength	dry	8.82 – 14.99 cN/tex (1.0 – 1.7 gf/d)
	wet	7.06 – 14.11 cN/tex (0.8 – 1.6 gf/d)
elastic recovery/ elongation		99% recovery at 2% extension, extends 20% – 40% when dry and 20% – 70% or more when wet
resiliency		excellent
density		1.30 – 1.32g/cm^3
moisture absorption		13.6% – 16% of fiber weight at standard conditions; 29% or more at saturation point
dimensional stability		subject to felting and relaxation shrinkage
resistant to	acids	good resistance to dilute acids; medium to poor resistance to concentrated acids
	alkalis	low resistance; many alkalis destroy fiber
	sunlight	Prolonged exposure deteriorates fiber
	organic solvents	good
	microorganisms	resistance generally good
	insects	damaged by insects such as moths and carpet beetles
thermal reactions	to heat	avoid prolonged exposure to temperatures above 130℃
	to flame	burn slowly when in direct flame; considered to be self-extinguishing

First, wool has a dry and warm hand because it is protein material, and it is light (low specific gravity). Next, it is very resilient, so wool fabrics insulate well; this, along with great loft, allows warmth with little weight. The resilience plus good elastic recovery mean wrinkle recovery. Wool can have a luster, but the finer the scales, the duller the fiber. However, when fine fibers are spun into worsted (combed) yarns, with fibers laid nearly parallel, we see the quiet luster evident in fine suitings.

折皱回复
精纺毛纱

Wool has low tenacity (tensile strength): pilling is not a great problem in wool fabrics. Pills can be removed with a stiff bristle brush, and so are not classed as "stubborn". However, wool does make very durable fabrics, with good abrasion resistance, which can be restored to good looks over a long wear life.

起毛

All hair fibers (and leather), since they are protein, can be molded or shaped with steam. A good example is the shaping of wool felt hats. This is also used in tailoring to give a temporary set, and is an important adjunct to shaping apparel in hard tailoring, where interlinings such as tailor's or hair canvas hold the structured shape of a jacket or coat. It is, of course, lost if a cheaper tailor's canvas of tape-like rayon fiber (cellulose) is used instead of a wiry hair fiber and wool.

Wool has natural flame-resistance unless humidity is very low or the fabric is very open. The coarser the wool, the more prickly it will be; this can cause itching and redness that is often taken for an allergic reaction. Since wool is the same protein as our own hair, a truly allergic reaction is rare, and most people can wear fabrics made of fine wool in finer, smoother yarns without discomfort.

Felting of wool described under Wool Structure is looked on as a disadvantage insofar as it is produced by machine washing and drying of wool articles with resulting matting and shrinkage, unless they have been given an antifelt treatment. However, it is often done on purpose, e. g., to make felt. In finishing, gradual felting is called fulling, giving closer, warmer fabrics—a definite advantage in uses such as winter coats.

毛毡
缩绒

2.2.2.3 Other Fur Fibers

其他毛皮纤维

Fibers from animals other than sheep (as shown in Fig. 2-20) Camels, Angora goat, Llamas, Alpaca, Vicuna, Musk oxen frequently are referred to as specialty fibers. Most specialty fibers come from members of the goat and camel families. Only those fibers from animals usually used for their pelts are labeled as fur fibers; these include mink, chinchilla, muskrat, nutria, raccoon, and rabbit. Both specialty and fur fibers are similar to sheep's wool in chemical and physical properties. They are added to wool fabrics primarily to give softness, color interest, and prestige value.

骆驼
无峰驼/羊驼/骆马/麝鹿/特种动物纤维
毛皮
貂皮/南美栗鼠皮/浣熊皮

(1) Camel Hair

驼毛

Camel hair is obtained from the two-humped Bactrian camel [see Fig. 2-20 (a)] of Asia. The fiber is gathered in the molting period. The coarse guardhairs separated by hand from the fine, soft undercoat. Fine camel hair products require the same careful handling and care as cashmere and mohair. Camel hair is used in coating fabrics, sportswear, and knitted products. The natural tan to reddish brown color is very attractive and is retained for many choice fabrics. Coarse camel

双峰骆驼
绒毛

hair fibers are used in industry for special belting, ropes, and artist's brushes. Natives in various areas use the coarse fibers to make blankets.

Since camel hair possesses thermal properties similar to those of wool, garments made of it keep wearers warm in cold weather.

(a) Camels (b) Angora goat (c) Llamas (d) Alpaca (e) Vicuna (f) Musk oxen

Fig. 2 – 20 Animals other than sheep

(2) Mohair 马海毛

Mohair comes from the Angora goat[see Fig. 2. 20 (b)], thought to originate in Asia; the name for the goat comes from Ankara, Turkey, also (confusingly) the origin of the name of the rabbit (with hair called angora). Mohair grows very long, up to 20 – 30 cm, if the animal is sheared once a year; but shearing is usually done twice annually, and the fibers are 10 – 15 cm in length. 安哥拉山羊

The fibers are fine and silky in appearance. Mohair is similar to wool in both physical and chemical properties. Its major advantages include remarkable resistance to wear and abrasion, a high degree of luster, excellent resiliency, and adaptability to complex yarns and textured fabrics. Suiting and sportswear fabrics, upholstery, rugs, and draperies may be of mohair or of a blend including mohair fibers.

(3) Cashmere 山羊绒

Cashmere is the underhair of the cashmere (Kashmir) goat, combed out as the animal molts. The short fibers (2. 5 – 8. 9 cm) are very soft and fine, used in sweaters; longer fibers (5. 0 – 12. 7cm) are somewhat coarse and stiff, used in overcoats. 换毛 运动衫,毛衣

The yield per goat averages about 114g of good fiber and the yearly production of true cashmere is very small, so the fiber is expensive. Fabrics made of cashmere are warm and comfortable. Furthermore, the fiber is highly adaptable, in that it can produce either fine or thick yarns, which, in turn, can be constructed into thick, medium, or rather lightweight fabrics. Cashmere is found in knits for sweaters and suits and in woven fabrics for apparel such as suits, coats, and dresses.

化学纤维 2.3 Man-made or Manufactured Fibers

纺丝的基本方法 2.3.1 Basic Spinning Process

Man-made or manufactured fibers (MF) are polymeric forms that are produced by some type of chemical action or by the regeneration of natural polymers in a new physical form. After the polymers are formed, the chemicals are converted into a liquid or fluid state and forced through a spinnerette (喷丝头) (Fig. 2 - 21). Spinnerettes with extrusion holes of shapes other than circular may be used for some fiber forms. The silkworm was the direct inspiration for MF, since the process could be seen and is simple: a sticky liquid is ejected in a thin stream, which turns into a solid as a filament.

There are some basic processes used in the manufacture of man-made fibers: wet spinning, dry spinning, melt spinning, solvent spinning, emulsion spinning (乳液纺丝) and gel spinning. The first MF were made using a naturally formed polymer, cellulose, as the raw material.

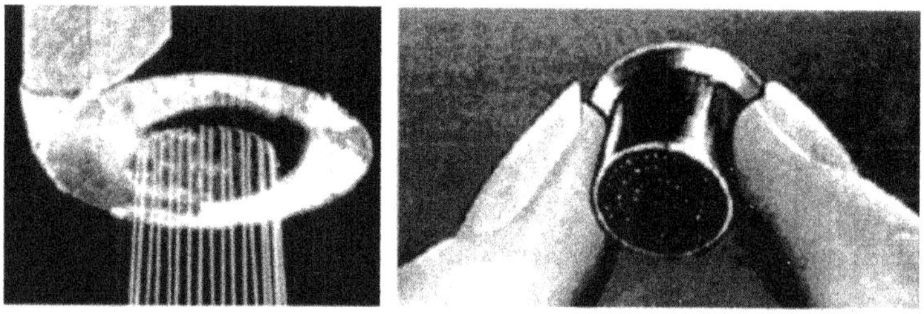

(a) Spinning filament fibers through a spinnerette

(b) The spinnerette that forms the spinning solution into filaments

Fig. 2 - 21 Spinnerette

湿法纺丝 (1) Wet Spinning

The fiber-forming liquid is extruded into a bath of liquid, in which the filament is solidified (固化). When the chemical solution is forced through the spinnerette, it emerges into a liquid that causes coagulation (凝固) and filament formation. This process may involve conversion of the fiber raw materials into a soluble derivative (可溶性衍生物); this derivative is extruded through the spinnerette into the coagulating bath (凝固浴), where additional chemical reactions convert the material back into the desired polymeric form. This process is called regeneration (再生). Fig. 2 - 22 is a diagram of the wet spinning process. Rayon and acrylic fibers are produced by wet spinning.

干法纺丝 (2) Dry Spinning

In dry spinning, a solid raw material is dissolve in a solvent that evaporates,

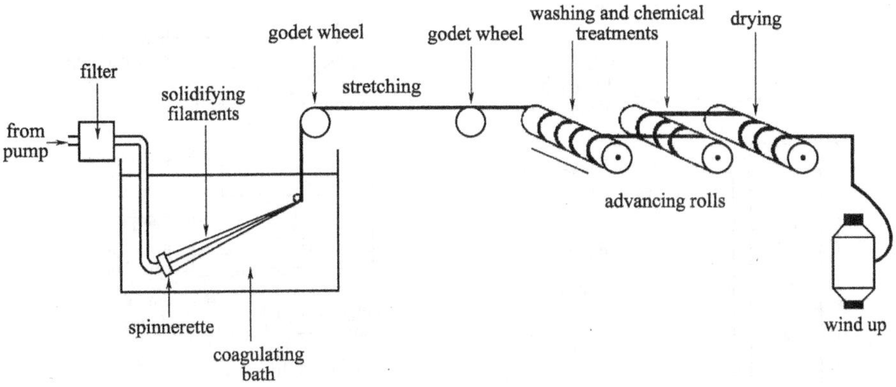

Fig. 2 – 22 The wet spinning process

leaving the solid filament. The chemical solution is forced through the spinnerette and into a warm air chamber. The warm air causes the solvent to evaporate and the fibers are formed and harden. The process may involve converting the polymer into a form soluble in a suitable liquid, too. As the solvent evaporates, the fiber polymer is reconstituted and returns to its standard chemical form but in a fibrous shape. Several fibers are made using dry spinning techniques (Fig. 2 – 23). Acetate and triacetate, some modifications of acrylics and modacrylics, and some aramid fibers are dry spun.

热空气通道

改性丙烯腈纤维/变性聚丙烯腈纤维/芳族聚酰胺纤维

(3) Melt Spinning 熔体纺丝

In melt spinning, a solid raw material is melted and spun, and forms the solid filament as it cools. The fiber polymer is melted and the molten solution is forced through the spinnerette. As the soft filaments emerge into a cooler environment, they harden into the standard filament fiber form (Fig. 2 – 24). Melt spinning requires no chemical change of any kind in the polymeric material from which the fiber is formed. Nylon, polyester, and glass are examples of melt spun fibers. 玻璃纤维

(4) Solvent spinning 溶剂纺丝

The fiber-forming material is dissolved in a solvent, then extruded and coagulated in a liquid bath with no chemical combination intermediate; an example is lyocell. 凝固

(5) Emulsion Spinning 乳液纺丝

Some raw polymeric chemicals cannot be processed by the processing methods mentioned above because they break down when heated to a melting temperature or they are not soluble in solutions that can be used. For these substances the emulsion process is applicable. The polymer is dispersed or emulsified into a solution, the dispersion or emulsion is forced through the spinnerette, and as the emulsion or dispersion leaves the spinnerette the fiber polymer forms into a fibrous shape. Depending on the type of fiber, the fibrous form produced by this method may be of staple or filament length. Teflon is an example of a fiber spun by the emulsion method.

特氟纶,聚四氟乙烯纤维

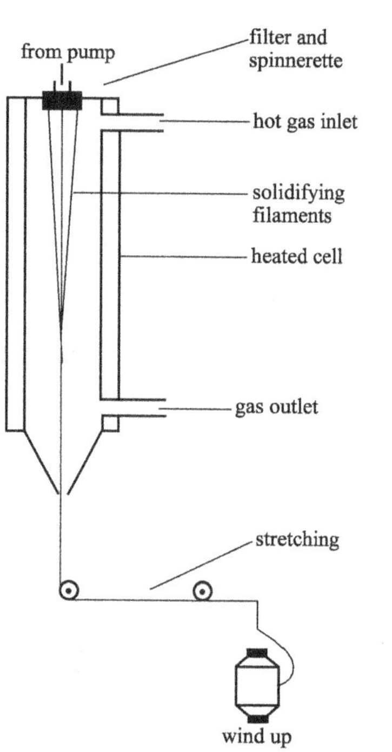

Fig. 2 – 23 The dry spinning process

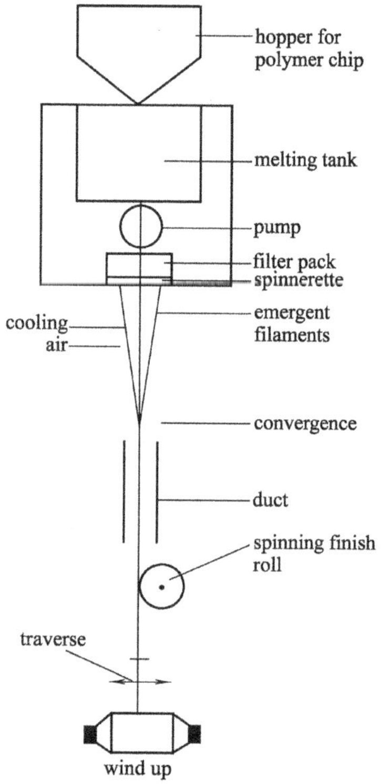

Fig. 2 – 24 The melt spinning process

凝胶纺丝 (6) Gel spinning

The fiber-forming material, as a very thick (viscous) liquid, is extruded, the filament is set as a gel, and the solvent washed off after; an example is the very high molecular weight high-tenacity fiber Spectra.

超高分子量聚乙烯纤维(商名,美国联合公司)

再生纤维

2.3.2 Regenerated Fibers

Of those MF fibers using natural polymers for raw material, some are minor, such as fibers from proteins, seaweed, or chitin; most of our discussion of this type will be of the cellulosic MF (CMF) fibers, which use cellulose as the raw material.

海藻/壳质

It can be noted that until 1953 in the United States all CMF fibers were classified as rayon. After that (much later in Canada), rayon was reserved for fibers regenerated by the viscose or cuprammonium process, while cellulose acetate compounds were given a separate generic grouping. In the 1990s, a new type of reconstituted cellulose fiber, Lyocell, was introduced commercially. It is produced by reconstituting cellulose from a solvent, using chemicals and processing methods that give superior CMF fibers and have little impact on the environment.

铜铵

莱塞尔纤维

2.3.2.1 Rayon

人造纤维,人造丝

Rayon, the oldest MF textile fiber has accounted for the greatest part of world CMF fiber production.

The Textile Fiber Products Identification Act defines rayon as:

A manufactured fiber composed of regenerated cellulose, as well as manufactured fibers composed of regenerated cellulose in which substituents have replaced not more than 15 percent of the hydrogen of the hydroxyl groups. 羟基

(1) Viscose Process 黏胶加工法

The principal raw material for viscose rayon is wood pulp. Cotton linters (the cotton fibers that are too short for yarn or fabric manufacturing) are also used. The wood pulp and linters are processed, and pure cellulose is extracted and formed into thin sheets that are about 0.6m^2. Considered chemically, the raw material is identical to pure cotton cellulose or to the cellulose in any of the vegetable fibers. 黏胶人造丝,黏胶人造纤维/木浆

Sheets of cellulose pulp are steeped in a 17.5 percent solution of sodium hydroxide until the cellulose is converted to alkali cellulose. 纤维素浆/氢氧化钠 碱纤维素

The alkalized sheets are then shredded into crumbs and aged for varying periods of time up to three days. After aging, the alkali cellulose crumbs are reacted with carbon disulfide. Up to this point the soda cellulose has been white, but the carbon disulfide changes it to a bright orange product called sodium cellulose xanthate. 熟成,老化 二硫化碳/碱纤维素 纤维素黄原酸钠

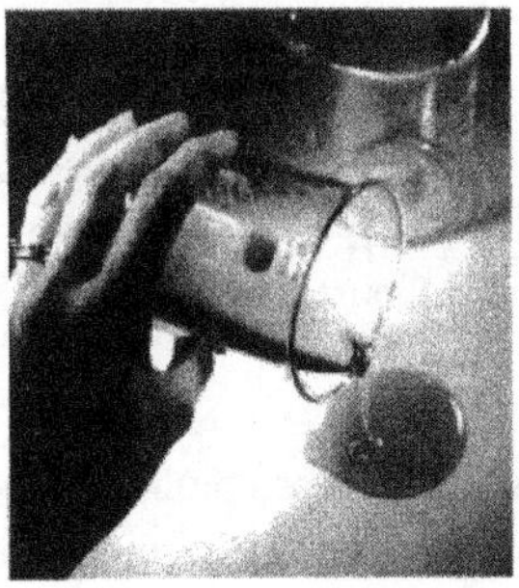

Fig. 2 – 25 A viscose solution of xanthate crumb dissolved in sodium hydroxide

The xanthate crumb is dissolved in dilute sodium hydroxide (5 to 8 percent NaOH) and forms a viscous (hence the name viscose), honey-colored liquid (Fig. 2 – 25). This solution is aged until it reaches the correct degree of coagulation for spinnability at specified conditions. The viscose is filtered and deaerated, delivered to the spinning pumps, and forced, by individual pumps for each spinnerette, into an acid bath. The bath and spinning conditions for viscose rayon are controlled carefully to maintain fiber quality. 脱泡

Fig. 2 – 26 is the basic flow diagram for viscose rayon production.

(2) Modified Viscose Fibers 改性黏胶纤维

The viscose process can be modified to leave the cellulose chains longer, with more orientation, giving an high wet-modulus (HWM) rayon (the generic name is modal in a number of countries) that comes closer to cotton in strength. Its strength when dry is fairly good, better than standard rayon, and it loses less strength when wet. 高湿模量 莫代尔

The higher wet-modulus means that the fiber is less prone to stretch when damp or wet; heavy drapes of standard rayon, for instance, lengthen in high humidity and shorten in drier air—what is called an "elevator effect". Otherwise, HWM rayons behave like any rayon. The main trademark name for HWM rayons is

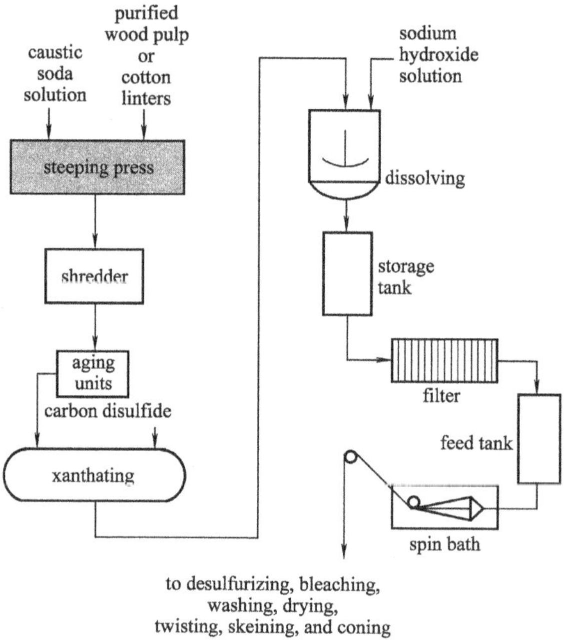

Fig. 2 – 26 Flow chart for viscose rayon manufanture

Zantrel ® (by BASF), with Lenzing Modal using the non-United States generic
波里诺西克,富强纤维 name modal. There is also an unofficial term, polynosic, for one type of HWM rayon. Avril (by American Viscose) and Vincel (by Courtaulds) are no longer produced.

铜氨丝,铜氨纤维 (3) Cuprammonium Rayon

Cuprammonium rayon (generic name is cupro in a number of countries) is made by a process that is much simpler than that for viscose. Cuprammonium, a deep
氨水 blue solution of copper in ammonia, converts the cellulose to a liquid form believed to be a complex formed with cellulose by the cuprammonium solution. Cellulose from purified wood pulp is combined with the cuprammonium solution, and when the resulting sticky liquid is spun and washed with water, cellulose is regenerated as filaments. Since spinning takes place downward through a funnel, the fibers are stretched somewhat, making them finer and slightly stronger than those from the standard viscose process. Cuprammonium rayon is most often used in high-quality
衬里 lining fabrics, since it is made in a finer filament than most viscose. Cuprammonium does not regenerate slowly as viscose does, and so does not show a wrinkled surface; it has a round cross section.

形态结构 (4) Mophology

The length or longitudinal appearance of regular viscose rayon exhibits uniform diameter and interior parallel lines called striations. The cross section of the fiber
锯齿形 shows highly irregular or serrated edges (Fig. 2 – 27).

Fig. 2 -27 SEM of rayon, regular viscose

High-tenacity viscose is similar to regular viscose when viewed under magnification, except that it may have less irregular contour in cross section and, thus, show fewer striations in the longitudinal view. High wet-modulus viscose rayon may appear nearly round in cross section and, therefore, unstriated in the microscopic longitudinal view.

In the longitudinal view cuprammonium rayon is uniform in width; it is smooth surfaced, and there are no markings or striations. The cross section is round or oval. It is relatively clear, and there are no irregularities in the contour.

(5) Properties 性能

The properties of rayon fibers are summarized in Table 2 -7.

Table 2 -7 Properties of Rayon Fibers

Property		Viscose			High Wet Modulus	Cuprammonium
		Regular	Medium	High		
shape		Shape can be controlled by the manufacturer. Therefore, it is uniform in appearance. Length and denier are determined by the manufacturer for the desired end use. Available in both filament and staple lengths. Diameter varies from 12 - 100μm.				
luster		Can be controlled by the manufacturer—from dull to bright				
strength in gpd	dry (gf/d)	1.5 -2.4	2.4 -3.2	3.0 -5.0	3.4 -5.5	1.7 -2.3
	wet (gf/d)	0.7 -1.4	1.2 -1.9	1.9 -4.3	2.2 -4.0	0.9 -1.3
elastic recovery at 2% extension		82%	97%	70% -100%	95%	75%
elongation	dry(%)	15 -30	15 -20	9 -26	6.5 -18	10 -17
	wet(%)	20 -40	17 -30	14 -34	7.0 -33	17 -33
resiliency		low			low-medium	low
density		1.50 -1.52g/cm^3 for all				

续表

Property		Viscose Regular	Viscose Medium	Viscose High	High Wet Modulus	Cuprammonium
sunlight		average for regular	medium	high	good	
dimensional stability		poor			good	
moisture absorption 20℃/65% RH		10.7% –16.0% (for all viscose types)				12.5%
saturation		25% –27% (for all viscose types)				27.0%
resistant to	acids	Resistance to concentrated acids is poor; to cold dilute acids, good.				
	alkalis	Resistance to strong alkalis is poor; to weak, good.				
	microorganisms	Mildew will destroy the fiber.				
	insects	Silverfish will damage all types of rayons.				
thermal reactions	to heat	Exposure to high temperatures eventually degrades the fiber.				
	to flame	All types of rayon burn readily				

莱塞尔 **2.3.2.2 Lyocell**

Lyocell is a solvent-spun cellulosic fiber—an innovation of the 1990s in this oldest group of manufactured fibers. Research went on at Courtaulds Fibres Plc. U. K.
样品试制阶段 from the late 1970s, with the pilot plant stage reached in 1988, and first commer-
天丝 cial production in 1992. This was staple fiber, with the trademark of Tencel®
(produced first by Courtaulds, now by Acordis; for logo, see Fig. 2 –28). Tencel
聚酯,涤纶 is used alone and blended with cotton, flax, polyester, and wool, plus some
弹性纤维,氨纶 blends with spandex. The first filament lyocell, trademark name NewCell® (by Acordis), offers wider applications of this fiber type in "silky" fabrics.

Lyocell as a generic name was established promptly in Europe, where the generic rayon has not been used since the 1970s, and was later accepted as a generic in the United States. In Canada it can be used as a generic or classified as another rayon.

加工和形态结构 (1) Process and Mophology

Lyocell is made by dissolving purified cellulose from wood pulp in an amine oxide. N-methylmorpholine-oxide (NMMO), an organic solvent that is much less harmful or irritating than others used in fiber production. After the clear, viscous (thick, sticky) liquid is filtered, it is extruded (spun) in a bath of dilute

Fig. 2 – 28 Tencel logo

NMMO, where the fiber coagulates as reconstituted cellulose. It is then washed and dried. Like cuprammonium, lyocell has a smooth surface and a round cross section. Since this is a closed-loop system, and virtually all the solvent can be recovered and used again, the process is largely pollution-free and is estimated to take only one-sixth of the time needed for the viscose process. A similar process was developed by Lenzing in Austria (Lenzing Lyocell), and others are now made, e. g., Junlon® (from Japan), and Turel™ (by Tectonic).

闭环系统 无污染

(2) Properities 性能

Lyocell has dry strength comparable to cotton and some polyester, very high wet-modulus, and it retains 85 percent of its dry strength when wet, so it will stand up well to home laundering. It can also withstand wet mill finishing to give durable-press, for instance, so wrinkling and poor wet press retention can be counteracted. Of course, being made up of cellulose, it has the advantages of good moisture absorption, good static resistance, cool hand, and good heat resistance.

Lyocell fibers can have a tendency to fibrillation— to splitting off if abraded when wet, forming surface fibrils (see Fig. 2 – 29). These give a very soft "peach skin" hand, which can also be promoted in finishing by physical abrasion, by enzyme treatment, or by chemical (alkali) treatment. In making some industrial fabrics, a "fast fibrillating" type is useful to generate microfibrils, as for filters. However, the tendency to fibrillation can also result in a difference in light reflection from the surface that may be interpreted as graying. As well, without fibrilla-

原纤化 桃皮绒效果 酶处理

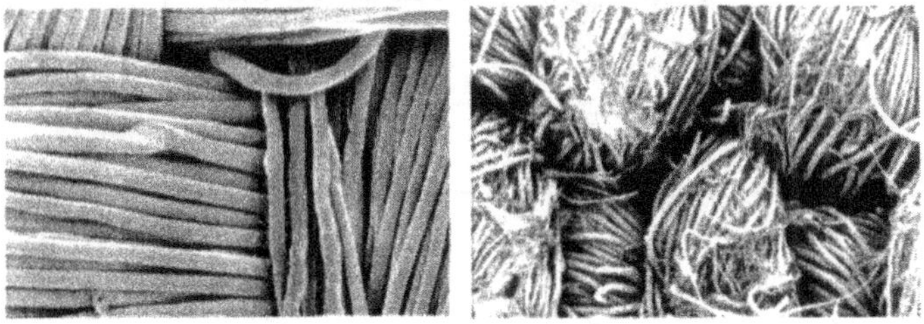

(a) Before fibrillation (b) After fibrillation

Fig. 2 – 29 Fabric of Tencel® lyocell

tion, different hands can be achieved that are desirable for certain products; there- 非原纤化的 fore a non-fibrillating variation is available. Tencel® A100 (by Acordis), and Lyocell LF (by Lenzing).

醋酯与三醋酯纤维 2.3.2.3 Acetate and Triacetate Fibers

Cellulose will combine with acetic acid to produce new compounds with very different properties from the original cellulose, and therefore different from natural cellulosic and reconstituted cellulose MF fibers.

三醋酯纤维 Cellulose acetate compounds of two types are made into fibers. Triacetate is the one discovered first but not commercially made until, as Arnel, it was marketed in 1954 by American Celanese. Production of triacetate was phased out in North America in 1987 because the solvent used was considered environmentally unacceptable; it was still produced under the trademark names Arnel in Belgium and Tricel in Italy (by Novaceta), and also made in Japan, but has faded with the 二醋酯纤维 1990s. The generic called simply acetate, introduced about 1905, is a secondary 二醋酯纤维 cellulose acetate compound, sometimes called a "diacetate"; in Europe, a trademark name for acetate was Dicel®.

生产 (1) Production

Cellulose is obtained from either wood pulp or cotton linters. The steps in the manufacture of acetate fiber are illustrated in Fig. 2-30.

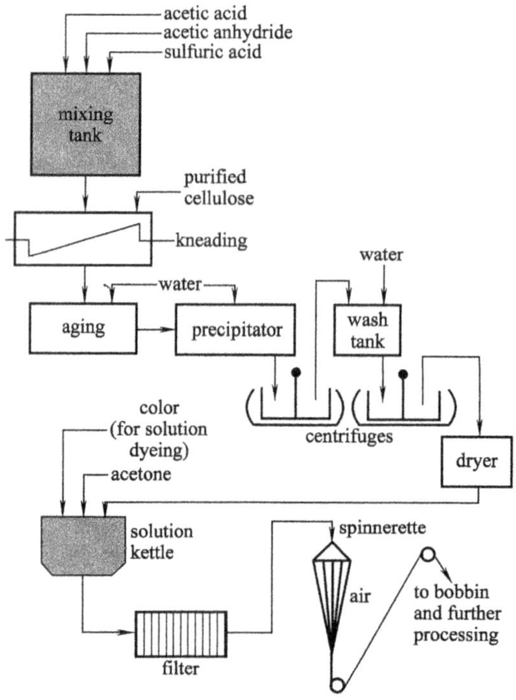

Fig. 2-30 Flow diagram showing processes in acetate manufacture

(2) Morphology 形态结构

When viewed longitudinally, acetate is uniform in width, with several lines parallel to the length [Fig. 2 – 31 (a)]. The cross section of acetate is lobed in outline with irregular curves but no sharp serrations like those found in viscose [Fig. 2 – 31 (b)].

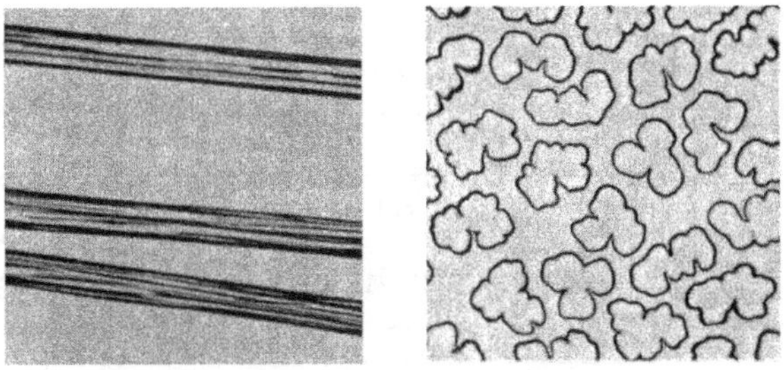

(a) Longitudinal view (b) Cross section view

Fig. 2 – 31 Photomicrograph of acetate

Triacetate is very similar in microscopic appearance to acetate [Fig. 2 – 32 (a)]. It may have clearer striations, and the cross section is both lobed and somewhat serrated [Fig. 2 – 32 (b)]. Positive distinctions between triacetate and secondary acetate cannot be made by microscopy.

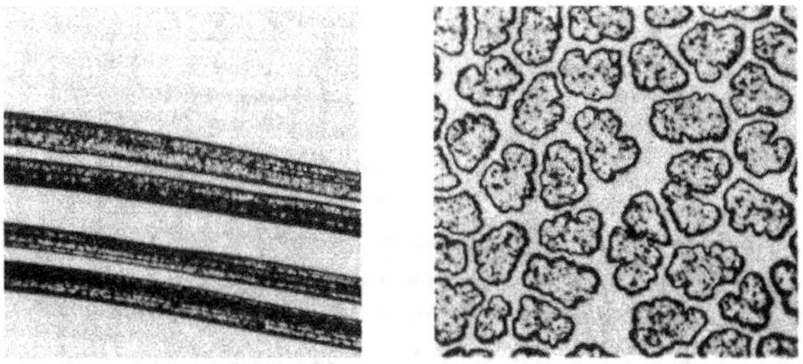

(a) Longitudinal view (b) Cross section view

Fig. 2 – 32 Photomicrograph of triacetate

(3) Properties 性能

Acetate and Triacetate fibers show both similarities and differences (See Table 2 – 8), since they are closely related, yet individual—like family members.

Cellulose acetate and triacetate are thermoplastic—they soften and melt on application of heat. Both are weak fibers with low abrasion resistance and tenacity, and both lose much of this low strength when wet. In burning, both fibers have an odor of vinegar, which relates to their composition; vinegar is weak acetic acid.

Triacetate softens at a higher temperature than acetate and can be given a heat-set in garment manufacture, using carefully controlled presses, that is permanent for the life of the article in normal use and care; pleats, for instance, will stay 热定形

sharp through washing and drying. It also withstands a fairly high iron heat, although care must be taken with any thermoplastic. It is a stiffer fiber than acetate, although with a softer hand than, for instance, nylon.

聚酰胺,尼龙,锦纶

Table 2 – 8 Properties of Acetate and Triacetate Fibers

Property		Acetate	Triacetate
shape		Shape can be controlled by the manufacturer. Therefore, it is uniform in appearance. Length and count are determined by the manufacturer	
luster		controlled for both; varies from dull to bright	
strength	dry	10.58 – 13.23 cN/tex (1.2 – 1.5gf/d)	9.70 – 13.23 cN/tex (1.1 – 1.5gf/d)
	wet	0.9 – 1.0	0.8 – 1.0
elastic recovery		94% at 2% extension 48% – 65% at 4% extension	92% at 2% extension 80% – 84% at 4% extension
elongation	dry	23% – 45%	25% – 40%
	wet	35% – 45%	30% – 40%
resiliency		low	good
density		1.32g/cm^3	1.30g/cm^3
moisture absorption	stand conditions	6.5%	3.2% – 3.5%
	saturation	14.0%	9.0%
dimensional stability		good	good
resistant to	acids	Both fibers have poor resistance to dilute acids, and are destroyed by most concentrated acids	
	alkalis	Both fibers have good resistance to dilute alkalis, both are eventually destroyed in concentrated alkalis	
	sunlight	loss of strength for acetate; triacetate has superior resistance	
	microorganisms	Both discolor; acetate will lose strength but triacetate retains strength	
	insects	Silverfish will eat sized acetate or triacetate	
thermal reactions	to heat	softens and melts at temperatures above 175℃ (350 ℉)	softens at temperatures above 235℃ (455 ℉)
	to flame	Both fibers burn easily and quickly	

稀有的天然高聚物再生纤维

2.3.2.4 Minor Natural Polymer Manufactured Fibers

(1) Alginate Fiber

海藻纤维、藻朊酸

Alginate fiber is made from alginic acid, derived from seaweed, and converted into calcium alginate to form a filament. This fiber is used because it will dissolve in an alkaline solution, such as of washing soda (sodium carbonate) and can be used for sewing or looping articles together temporarily or as a ground material that is to be removed eventually. It has also found use in medicine, since alginate will help stop bleeding and does not irritate.

藻酸盐/褐藻酸
藻酸钙
洗涤碱/碳酸钠

(2) Chitin Fiber

甲壳素纤维

Chitin is a polymer based on a sugar-like unit that resembles cellulose. It is the main part of the hard outer covering of insects, arachnids, and crustaceans, as well as forming reinforcement in the cell walls of fungi, molds, and yeast. As such it is the most abundant organic compound after cellulose. It can be made into a fiber; in medicine it is absorbable as sutures and has been found to speed healing greatly when used as wound dressings. Its derivative chitosan may also be used to remove offensive color from dyebath effluent, in combating water pollution.

增强，加固
缝合线
创伤敷料/壳聚糖

(3) Azlon (Protein)

阿泽纶(人造蛋白纤维)

Azlon fibers are man-made protein fibers. The TFPIA defines azlon as:

A manufactured fiber in which the fiber-forming substance is composed of any regenerated naturally occurring protein.

To make azlon fiber, protein is extracted from a source such as casein, peanuts, or corn. The protein is processed into a spinning solution, extruded into a coagulating bath, and filaments are formed.

酪蛋白/花生
玉米/凝固浴

In general, protein fibers are relatively weak, 7.9 – 13.2 cN/tex (0.9 – 1.5 grams per denier) in tenacity. Specific gravity varies from 1.25 to 1.3g/m^3. The fiber has good elastic recovery and elongation, high moisture absorbency, and superior resistance to microorganisms and insects. Azlon fibers are exceptionally soft and make fabrics similar to soft wool or cashmere. Negative aspects include the fact that the fibers have an unpleasant odor when wet.

Azlon fibers are used in blends with stronger fibers. They add softness and warmth to fabrics, and when blended with wool they reduce the tendency for the fabric to shrink.

2.3.3 Synthetic Manufactured Fibers

合成纤维

Synthetics have been made using fossil fuels (i. e., oil, gas, coal) plus air, water, and other chemicals as the raw materials. From these, chemicals are obtained to make the simple materials that will be the units of the fiber polymer. These units must have reactive groups at both ends; in polymerization, these groups join to form very long chain molecules with fiber-forming properties. Since the beginning of the 21st century, synthetics are being formed also with units made from renewable materials, such as corn starch.

反应基
玉米淀粉

2.3.3.1 Nylon

锦纶

Nylon was first marketed in 1938 as bristles and fish lines, and in the early

针织品,袜类 1940s as nylon hosiery or "nylons".

Nylon contains carbon, hydrogen, oxygen, and nitrogen. The first nylon was and is referred to as type 66. Nylon 66 gets its name or identification from the fact that 己二胺/脂肪酸 it is made of two basic chemicals, hexamethylene diamine and adipic acid, each containing six carbon atoms per individual molecule. Nylon type 6,10— composed of one chemical with six carbon atoms and one with ten—was developed simultaneously, and DuPont used it in making bristles for toothbrushes, paintbrushes, and similar products.

Since the middle 1940s many other types of nylon—such as nylon 3; 4; 6; 7; 芳香的/脂肪族的 9; 11; 12; 13; 6,13; and aromatic, aliphatic combinations—have been produced. In 1973 the Tederal Trade Commission rules and regulations for fiber identification established a new generic term, aramid, to identify aromatic polyamides. 芳香族聚酰胺/芳香锦纶

The TFPIA defines two polyamide fibers nylon and aramid as:

Nylon—a manufactured fiber in which the fiber-forming substance is a long-chain synthetic polymer, in which less than 85% of the amide linkages attach directly to two aromatic rings.

Aramid—a manufactured fiber in which the fiber-forming substance is a long-chain synthetic polyamide, in which at least 85% of the amide linkages attach directly to two aromatic rings.

生产 (1) Production

Nylon production starts with the basic chemicals for any one type of nylon (derived mainly from petrochemicals), combined to form the unit compound; then polymerization takes place under heat and pressure. The polymer, a thick, sticky liquid, is solidified and broken into white flakes or chips. The chips are melted and spun (see Fig. 2-33). Melt-spun fibers hold the shape of the spinnerette hole, so that modifications in cross section can be made. The resulting fiber is drawn to orient the molecular chains. (see Fig. 2-34). 石油化学制品

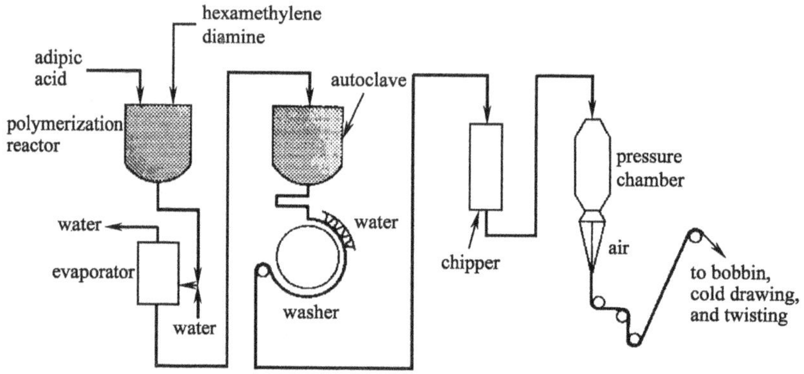

Fig. 2-33 Flow diagram showing processes in nylon 66 manufacture

形态结构 (2) Morphology

Nylon filaments are smooth and shiny. When viewed in cross section, nylon is

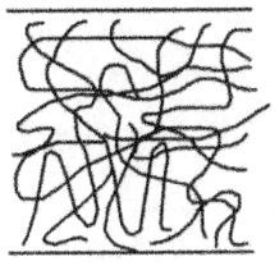

(a) Undrawn polymer; no orientation or crystallinity (amorphous)

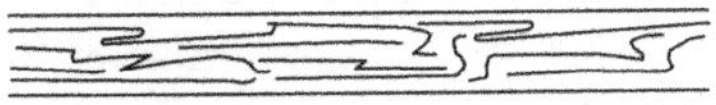

(b) 3 : 1 draw ratio; alignment of some chains and crystallinity introduced

(c) 6 : 1 draw ratio; high degree of chain orientation and crystallinity

Fig. 2 – 34 Effect of drawing on molecular orientation

usually perfectly round. Exceptions to this are trilobal nylons such as Antron, Cadon, Cumuloft, and Qiana. Longitudinal magnification shows relatively transparent fibers of uniform diameter, with a slightly speckled appearance. 三叶形

(3) Properties 性能

Nylon's properties are outstanding; although no longer the most used synthetic, nylon is still the strongest major fiber in its standard form. It was first used as filament in place of silk to give strong, light materials such as parachutes and women's sheer hosiery. 降落伞

The properties of nylon are shown in Table 2 – 9.

Table 2 – 9 Properties of Nylon 66, Nylon 6, and Qiana Fiber

Property		Nylon 66	Nylon 6	Qiana *
shape		Shape can be controlled by the manufacturer. Therefore, it is uniform in appearance. Length and denier are determined by the manufacturer for specified types of end-uses Fiber is available in filament or staple length		
luster		Luster is controllable, from bright to dull		
strength	dry	11.47 – 79.88cN/tex (1.3 – 9.0 gf/d)	30.87 – 79.38 cN/tex (3.5 – 9.0 gf/d)	26.46 – 30.87 cN/tex (3.0 – 3.3 gf/d)
	wet	35.28 – 67.03cN/tex (4.0 – 7.6 gf/d)	28.22 – 70.56cN/tex (3.2 – 8.0 gf/d)	26.46 – 30.87cN/tex (3.0 – 3.3 gf/d)
elastic recovery		100% at 4% extension for all types		
elongation	dry	19% – 40%	16% – 50%	30% – 35%
	wet	20% – 46%	18% – 55%	30% – 35%
resiliency		good	good	excellent
density		1.14g/cm^3	1.14g/cm^3	1.03g/cm^3

续表

Property		Nylon 66	Nylon 6	Qiana *
moisture absorption	stand conditions	3.8% -4.5%	3.5% -5.0%	2.0% -2.5%
	saturation	8.0%	8.5%	2.0% -2.5%
dimensional stability		excellent	excellent	excellent
resistant to	acids	All nylon has poor resistance to acids. Some mineral acids, as well as some organic acids, dissolve nylon fibers		
	alkalis	Resistance of all nylons is good		
	sunlight	Sunlight is destructive to nylon fibers unless special finishes have been added.		
	microorganisms	high		
	insects	resistant to most; may be damaged by ants, crickets, and roaches		
thermal reactions	to heat	All nylons will soften and eventually melt		
	ironing	150 - 175℃	150℃	185℃
	temperature to flame	Heat sensitive; shrink from flame, melt, drip, burn in open flame; considered to be self-extinguishing. Some finishes and dyestuffs may alter behavior in flame		

奎阿纳纤维

* Qiana fiber is a trade name for polyamide with trilobal cross section.

改性锦纶

(4) Modified Nylon

In addition to developing nylon fibers from new chemicals, fiber scientists have been actively involved in modifying the common forms of nylon, type 66 and type 6.

改变截面形状

① Modified Cross Section

多叶形 / 喷丝孔

Original nylon fibers were round in cross section; modified cross sections include triangular, irregular, trilobal, and other multilobal shapes. These changes are made possible by altering the shape, arrangement, and number of orifices in the spinning jet.

挺爽的/闪光效应

Modified cross sections can produce many desirable qualities, such as increased cover; a crisp, silklike, firm hand; reduced pilling; increased bulk; sparkle effects; and heightened resistance to soil. Trilobal nylon is available in various fiber deniers or diameters and is used in apparel or home furnishing fabrics.

锦纶变形丝

② Textured Nylon

不透明的

Nylon filaments can be texturized to add bulk and/or stretch by many methods. In addition, special manufacturing processes produce filaments that will develop a crimp during finishing operations. These producer-textured fibers require careful handling during fabric making, but the end product has characteristics appealing to consumers, such as good elongation and outstanding recovery from stretch. The fiber also exhibits good cover, so opaque, lightweight fabrics can be made.

双组分锦纶

③ Bicomponent Nylon

肯特兰斯,双组分锦纶单丝或复丝

Cantrece is a side-by-side bicomponent nylon; it is considered a third-generation nylon; and it is the major bicomponent made of polyamides. This fiber is composed of two nylon components with different shrinkage potentials. Upon exposure to heat a latent crimp is developed as one component shrinks more than the other. The fiber has superior fit retention and resilience, which makes cantrece highly desirable for women's hosiery. Other bicomponent nylons include Heterofil, a sheath core type made in England, and Kanebo nylon 22, a side-by-side type made in Japan. Heterofil is used in carpets and bonded fiber fabrics, whereas Kanebo nylon 22 is used primarily for stretch underwear and hosiery. 皮芯型

④ Cross-linking 交联

Nylon molecules can be cross-linked by adding a chemical characterized by reactive groups at both ends of the chemical molecule. This chemical reacts with either the terminal amine group or the terminal carboxyl group on the nylon molecule to hook adjacent fiber molecules together. Such modifications increase the fiber modulus and are especially useful in fibers intended for vehicle tires. 端氨基/端羧基

⑤ Graft Polymerization 接枝共聚

Nylon fibers can be modified by grafting other chemicals onto the fiber molecules. One important property altered by graft polymerization is moisture absorbency, which is increased, and in turn, increases the wet crease recovery and diminishes static buildup.

⑥ Antistatic Nylon 抗静电锦纶

Ultron nylon is one of the antistatic fibers made by the addition of a small amount of a conducting polymeric substance to the melt solution before spinning the filaments. This fiber has a silklike luster, good covering, power, resistance to soiling, and superior whiteness retention. Static properties are similar to cotton. It is used primarily for carpeting.

The fiber X-static is made by coating nylon filaments with a microscopic layer of metallic substances. Antistatic nylons may be made also by finishing processes.

Antron III is made with a polymeric conductive compound incorporated into the polymer which imparts permanent static control to yarns and fabrics.

⑦ Dye Behavior Modification 染色性能改性

Nylons that accept different types of dyestuffs are produced through the incorporation of small amounts of special chemicals into the melt before spinning. Nylons that are "deep dyeing" are used in carpets to produce special color patterns.

⑧ Reduction of Damage by Light 减少光损害

Antioxidant and light-absorbing chemical substances can be added to the melt before filament formation. These products reduce the yellowing and lose of strength that occurs in nylon after long periods of exposure to light and particularly ultraviolet rays.

⑨ Flame-retardant Nylon 阻燃锦纶

DuPont markets a fiber, Hyten, that has special substances added to increase the flame resistance. It is used in industrial situations where special flame resistance is needed.

2.3.3.2 Polyester 涤纶

Polyester was the second synthetic fiber type to be developed (1941) and is now by far the most widely used, due to a balance of properties that gives it great versatility plus a less expensive manufacturing process compared to nylon.

聚对苯二甲酸乙二酯/乙二醇 对苯二酸 There are several types of polyester. The original type is called PET for polyethylene terephthalate, the result of linking and then polymerizing ethylene glycol (antifreeze) and terephthalic acid.

Polyester is made up only of carbon, hydrogen, and oxygen (no nitrogen, as there is in nylon): the carbons in the acid are in rings, making this, in organic chemistry terms, an aromatic acid.

Some of the many trademark names for PET polyesters are Dacron® (by DuPont and DAK Americas); Diolen® (by TWO); Encron® (by BASF); Fortrel® (by Wellman); Kodel® (by Eastman); Tergal® (by CEP), Terital® (by Montefibre); Tetoron® (by Teijin, Toray); Trevira® (by Trevira); and Wistel® (by Snia). One of the large producers of commodity polyester is KoSa. which took over the North American business of Hoechst, and has various trademark names for specialty versions.

生产 (1) Production

Production of polyester follows steps similar to those outlined for nylon; it is also melt-spun and drawn after the fiber is made.

形态结构 (2) Morphology

The longitudinal view of polyester exhibits uniform diameter, smooth surface, and a rodlike appearance. The cross section of regular polyester is round. There are several variations in cross-sectional shapes. The most common include a trilobal variant of Dacron (Fig. 2 – 35), a T-shape variant of Fortrel (Fig. 2 – 36), a 五叶形 pentalobal variant form from Hoechst (Trevira Star), an eight-sided-star variant of Encron, and a "trilateral" variant of Kodel.

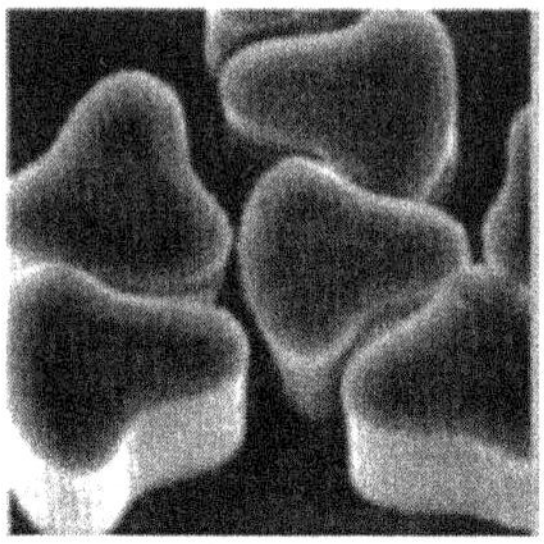

Fig. 2 – 35 Photomicrograph of trilobal polyester

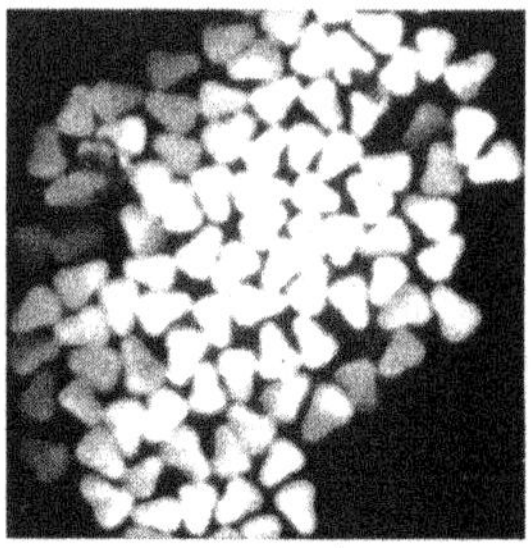

Fig. 2 – 36 Photomicrograph of T-shape polyester, cross section

性能 (3) Properties

Properties of polyester (See Table 2 – 10) can appear very similar to those of nylon, so it is important to clarify how they differ. One thing is certain: If a MF fiber does not have specific, significant advantages over others, it will not be, or continue to be, produced.

Table 2 - 10 Properties of Polyester Fibers

Fiber		Tenacity Wet and Dry (gf/d)	Elongation Wet and Dry (%)	Elastic Recovery at 2% Elongation (%)	Density/ g/cm^3	Moisture Regain at 20℃; 65% RH(%)
Dacron	regular	2.8 - 5.2	19 - 30	97	1.38	0.4 - 0.8
	high-strength	6.0 - 9.5	10 - 14	100	1.38	0.4 - 0.8
Kodel	Type Ⅱ	2.5 - 3.0	24 - 34	85 - 95	1.23	0.4
	Type Ⅳ	4.5 - 5.5	35 - 45	75 - 85	1.38	0.4
Fortrel	regular	3.75 - 4.88	25 - 30	100	1.38	0.4
	high-strength	7.0 - 8.5	8 - 10	100	1.38	0.4
Vycron	high-strength	5.8	34	90 - 98	1.38	0.6
	Type 2	3.8 - 4.4	49 - 58	93	1.38	0.6
	Type 5	3.9 - 4.9	43 - 67	98	1.38	0.6
Encron	regular	4.4 - 5.0	20 - 35	95 - 97	1.38	0.4
	high-strength	8.5 - 8.9	11.5 - 13.0	100	1.38	0.4
Avlin		3.0 - 4.6	18 - 22	90 - 97	1.38	0.4
Quintess		4.5 - 5.5	20 - 46	97	1.38	0.4
Monsanto		3.5	28 - 30	95 - 97	1.38	0.4
Trevira		3.1 - 5.8	20 - 63	95 - 97	1.34 - 1.38	0.4

Polyester has strength and abrasion resistance next to nylon and also takes a permanent heat set—all "first-class" fiber properties, from the point of view of performance. The further characteristics that finally put polyester ahead of nylon in amount used are good resilience (the nearest to wool of any MF fiber) and high modulus; these mean that it is springy and can recover well from strain, yet resists stretching better than nylon. Polyester also conducts moisture away from the skin better than nylon, dries faster, and is more supple in cold conditions.

Polyester does have particular drawbacks, too, one being that it holds oily stains tenaciously. It absorbs very little moisture, so it is less comfortable to wear in its standard form. It is also a static collector (of course, less absorbency means quick drying). Like nylon, polyester fabrics show stubborn pilling because of the great strength of the fiber.

2.3.3.3 Acrylic 腈纶

Acrylic fibers and polyester fibers were developed about the same time, and the commercial development of both was postponed by World War II. Dupont synthesis of an acrylic fiber evolved from the fundamental research of Dr. Carothers. Two distinct generic groups of fibers based on acrylonitrile chemistry are defined by the 丙烯腈

TFPIA:

Acrylic — a manufactured fiber in which the fiber-forming substance is any long-chain synthetic polymer composed of at least 85 percent weight of acrylonitrile units.

Modacrylic — a manufactured fiber in which the fiber-forming substance is any long-chain synthetic polymer comprised of less than 85 percent but at least 35 percent by weight of acrylonitrile units.

生产 (1) Production

丙烯氰 Acrylic fibers are made up of 85 percent of the unit acrylonitrile (vinyl cyanide). Production may be by wet or solvent spinning. One solvent for acrylonitrile 二甲基甲酰胺 is dimethylformamide.

形态结构 (2) Morphology

The cross-sectional shape of acrylic and modacrylic fibers is determined by the method used for spinning the fiber. The round cross section of Acrilan [Fig. 2 – 37 (a)] and Creslan [Fig. 2 – 37 (b)] suggests that they are spun into a liquid bath. Orlon [Fig. 2 – 37 (c)] has a dogbone shape, which could come from the evaporation of a solvent in dry spinning. The bicomponent fibers [Fig. 2 – 37 (d)] look like two fibers stuck together.

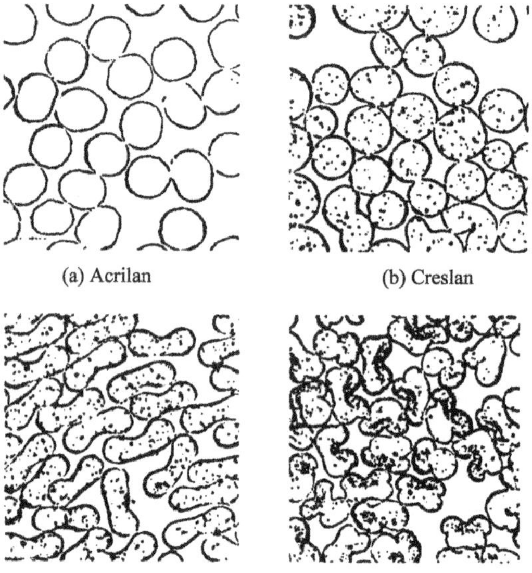

(a) Acrilan (b) Creslan

(c) Orlon (d) Bicomponent fibers

Fig. 2 – 37 Cross section of different acrylics

性能 (3) Properties

Acrylics produced by different companies can have a variety of characteristics, or one company can market differently dyeing acrylics.

The properties of acrylic fiber include the most wool-like hand of any MF: warm, dry, and soft. Acrylics are also lofty (of low density) with good resilience — a combination to make into wool-like fabrics that are warm with little weight. All acrylic in consumer use is staple fiber, and much is used in place of, or with, wool, since acrylic is much less expensive than wool, does not felt in machine washing, has fairly good strength, and does not lose strength when wet. Acrylic has taken over much of the market of wool in blankets and moderate- to lower-price clothing, especially for infants and children.

Acrylic has quite low absorbency, giving stain resistance and cleanability, as noted, but exhibiting static collection and lowering comfort in clothing.

Acrylic is more sensitive to heat than nylon or polyester — it is well to remember this when tumble drying articles of acrylic.

2.3.3.4 Polypropylene

丙纶

Polypropylene is the most usual member of olefin family in textile consumer goods. Olefins are paraffin-like (waxy), with rather simple units, which makes them the most economical synthetics to make. The other family member, polyethylene, is more often put to nontextile use in squeezable toys and kitchenware; it is sometimes used in textiles, for example as tent floors, as well as nonwovens.

聚丙烯/烯烃
蜡状
聚乙烯

(1) Production

生产

Most of the olefin fibers are melt-spun. The polymer is produced from the gas at temperatures of 55 to 65℃ at pressures of about 32 kg/cm^2, using a catalyst and a hydrocarbon solvent. The polymer is melt-spun at 300℃ into water or a current of cold air to quench the fiber. The filaments are cooled and cold drawn to about six times their spun length.

Polypropylene is produced not only by melt spinning but by the economical split-film method of fibrillating. The polymer is extruded in a sheet, which is then drawn and cut into narrow strips, or tapes.

(2) Morphology

形态结构

The shapes of olefin fibers may vary, but most are elliptical or round in cross section. Some fibers have an irregular, almost triangular shape. As with other manufactured fibers, they are available in a variety of lengths and count. Both lustrous and delustered forms are available.

(3) Properties

性能

Polypropylene has the lowest density of any fiber (specific gravity less than 1.0, lighter than water—like wax) and so is very lofty, giving good cover for low price. It has good tensile strength, abrasion resistance, wrinkle recovery, and chemical resistance. Its modulus is between that of nylon and polyester, and in spite of almost no moisture absorbency, it has good static resistance. All of the foregoing have ensured its use in carpets and home furnishing fabrics. However, again in spite of almost no moisture absorbency, olefin wicks, and so moved into use in garments, first as staple (spun yarn) in underwear, but later as filament in a wide variety of activewear. Innova® (by Amoco) has performed well in specialized sportswear such as for warm weather diving, paddlesports wetsuits, surfing, bike tights, and aerobicwear. Low absorbency gives olefin stain resistance in furnishing fabrics and quick drying in garments. With almost no absorbency, polypropylene is difficult to dye; solution dyeing is the most effective way to color the fiber.

氨纶

2.3.3.5 Spandex

Elastomers are elastic, rubber-like substances. They can be stretched to several times their length and will recover immediately after the tension is released. For many years the only elastomer was rubber, both natural latex (the sap of the rubber tree) and synthetic rubber.

天然乳胶 合成橡胶

The major synthetic elastomer, spandex, should be compared only to rubber in properties. It does not conform to the general behavior of synthetic fibers, since its strength is relatively low. Spandex can stretch up to seven times its original length and recover instantly when tension is released.

The first spandex was introduced by the DuPont Corporation in 1958. Plans for volume production and the trade name, Lycra (see Fig. 2 – 38), were announced to the public in late 1959.

莱卡

Fig. 2 – 38 Hangtag for Lycra®

生产

(1) Production

Spandex fibers may be produced by melt spinning, wet spinning, or solvent spinning. One process, used by globe, is described as a reaction spinning process in which an isocyanate-terminated pre-polymer is extruded into a diamine bath. It is likely that in the production of Lycra, the cured polymer is dissolved in a solvent such as dimethylformamide and extruded into a spinning column containing solvent vapor and inert gas. The fibers coagulate; then they are washed, dried, and wound on bobbins.

异氰酸盐，异氰酸酯

形态结构

(2) Morphology

Spandex fibers are usually sold as continuous filaments, although some staple fibers have been produced for blending with other fibers in spun yarns. In cross section Lycra has a round or dogbone shape: Glospan fibers look like fused, hollow-core, lightly serrated filaments.

中空

性能

(3) Properties

Spandex fibers are also fairly weak, with tenacities ranging from 4.4 to 13.2 cN/tex (0.5 to 1.5 gf/d). The elongation of spandex fibers ranges from 500 to 700 percent, with 100 percent recovery at 2 percent elongation. At 50 percent elongation, spandex has 99 percent recovery; at 200 percent elongation, it has 97 percent recovery. Its wet elastic recovery is similar to its dry elastic-recovery. The elastic modulus of spandex is higher than that of rubber. The moisture regain of spandex ranges from 0.3 to 1.2 percent for variants of the fibers. High temperatures adversely affect the elasticity of spandex fibers, but short exposures at temperatures of 95 to 150℃ do not damage the fiber. Spandex fibers are sensitive to high concentrations of chlorine. Spandex filaments are dyeable, so uncovered filaments may be used in fabric constructions and dyed to match other fibers in a blend.

2.3.3.6 Vinal, Chinon, Kevlar, ect.

维纶,蛋白质接枝聚丙烯腈纤维,凯夫拉等

The textile industry is constantly changing. The following fibers have been introduced for special end-uses, and most of them have selected characteristics that make them desirable for industrial applications. Some may be used for specialized consumer products.

(1) Vinal (Vinylal) 维纶

Vinal, also Given as PVA, Polyvinylalcohol, is made up of vinyl alcohol with various acetal units. As first made, this produces a fiber soluble in water, useful as a ground for lace or in blends to give sheer fabric when it is dissolved away, or as a connecting thread in knitting. Further treatment develops a strong fiber with good elongation that melts but is self-extinguishing, with good chemical and rot resistance.

聚乙烯醇纤维
自熄灭/防腐

Since vinyl alcohol is unstable, the polymer is formed indirectly by the hydrolysis of polyvinyl acetate. The fibers are water soluble when extruded and must be treated with formaldehyde to make them insoluble. The treatment produces intramolecular and intermolecular crosslinks that stabilize the polymer.

水解
聚乙酸乙烯酯
甲醛/分子内的
分子间的

(2) Chinon 蛋白质接枝聚丙烯腈纤维

Chinon is the trade name for a bicomponent of 70 percent acrylonitrile and 30 percent azlon produced by a graft polymerization process. The fiber is produced in Japan from milk casein and acrylic resins and is identified as a promix fiber.

乙烯系单体与蛋白质共聚纤维

(3) Kevlar 凯夫拉

Kevlar, a kind of aramid fiber, has an extremely high tenacity of approximately 198 cN/tex (22 grams per denier), which is more than five times the strength of steel wire and more than twice the strength of high-tenacity industrial nylon, polyester, or fiberglass. It has an unusually high initial-modulus, resistance to initial pull and stretch, of 475 grams. This is about two times greater than fiberglass or steel, four times greater than industrial polyester, and nine times greater than industrial nylon. Moisture regain is about 7 percent. Elongation is low, only 4 percent.

初始模量

Thermal properties of Kevlar are superior. It has outstanding stability to heat, retaining a high percentage of strength after exposure to temperatures to 260℃. Kevlar has superior resistance to most chemical reagents. Acids do cause a loss in strength, and hydrochloric and nitric completely degrade the fiber. Concentrated alkalis at elevated temperatures attack the fiber.

(4) PLA 聚乳酸

Poly lactic acid (PLA) has generated great interest as one of the most innovative materials being developed for a wide range of applications. The polymer is thermoplastic and biodegradable, which makes it highly attractive for biological and medical applications. It can be transformed by spinning into filaments for subsequent

聚乳酸
可生物降解的

fabrication of desirable textile structures. Spinning may be accomplished by various routes, each with its merits and demerits. The medical applications of this polymer arise from its biocompatibility: the degradation product, lactic acid, is metabolically innocuous. The fibers may be fabricated into various forms and may be used for implants and other surgical applications such as sutures. Tissue engineering is the most recent domain where poly (lactic acid) is being used and is found to be one of the most favorable matrix materials.

组织工程

基质材料

Lactic acid is produced commercially by fermentation, or via petrochemical feedstocks.

PLA has several advantages: it is biocompatible and biodegradable, and can be readily broken down thermally by hydrolysis. It is available from renewable agricultural resources. This latter circumstance also helps to improve the farm economy. There is reduction in carbon dioxide emissions in comparison with conventional petroleum based commodity plastics because it helps by fixation of significant quantities of carbon dioxide. The most important ability of PLA is that one can tailor its physical properties by material modifications. The polymer is relatively hard, with the glass transition temperature in the range 60 - 70℃ and melting at 170 - 180℃.

聚对苯二甲酸丙二醇酯

(5) PTT

Up to now, PET is still popular materials in the world. Although the mechanical properties of PET are good enough to warrant its use as fibers, films, and engineering plastics, there are some difficulties in processing PET because of its high melting temperature. The diacid group affects the mechanical property of polyester and its processibility depends on the type of the diol group.

二酸

二醇

丙二醇

In particular, PTT, synthesized using 1,3 - propandiol as a diol, is a high crystalline polymer. Its melting temperature is lower than that of PET by 25℃. Therefore, the processibility of PTT is superior to that of PET. Furthermore, the highly flexible PTT fibers are obtained as a result of its low initial modulus. The elasticity and dyeability of PTT are better than those of PET or PBT, which make it possible to use PTT as engineering plastics, films, carpets, and clothing materials. For these reasons, PTT is considered as the most promising candidate for a replacement of PET in the world. So they will undoubtedly become an important new family of fibers. PTT has attracted quick attention from both fiber and textile industries. It has some of the characteristics of both PET and Nylon 6. One such characteristic is its atmospheric dispersed dyeing with excellent colorfastness and natural stain resistance. No pH adjustment, carrier, or high-temperature dyeing machine is necessary for the disperse dyeing of PTT.

聚对苯二甲酸乙二酯/聚对苯二甲酸丁二酯

2.4 Modifications of Manufactured Fibers

化学纤维的改性

One of the great advantages of a manufactured fiber is the possibility of producing it in many shapes and forms. One major area of modification gives us special versions of the major synthetics that are more pleasing and/or more comfortable — in other words, closer to natural fibers in aesthetics, but maintaining excellent performance. Japan has given us a name for such new synthetic fibers—Shin gosen.

新合纤

Special modifications introduce a variety of altered properties. Changes that alter the cross section produce fibers with different hand and appearance. Chemical modifications result in fibers that may be highly flame-retardant; different in strength (which alters pilling behavior); more crush-resistant; and dye-selective.

2.4.1 Specialized Physical Modifications

特殊物理改性

(1) Nonround Cross Section Fiber

异形截面纤维

Cross-sectional shape is changed by altering the shape of the openings in the spinnerette. These are produced to alter the characteristics of fibers that have a standard form with a round cross section; this applies to the two main synthetics, nylon and polyester. Nonround or irregular cross section can give a shape with lobes, like a clover leaf (trilobal, pentalobal, octalobal, or often just lobal). (See Fig. 2 - 39 and Fig. 2 - 40). Fibers with cross sections that are trilobal, pentalobal, or star-shaped have different characteristics related to hand, drape, and appearance than round fibers. Fabrics made from polyesters of trilobal and pentalobal cross section are characterized by a silklike hand and appearance. These fibers may also be used in carpet yarns, where they give a good appearance and appear to soilless quickly. This latter characteristic is due to the tendency for dirt to lodge in the valleys between the lobes of the fiber. Other properties given by the lobal shape include a pleasing luster and surface sheen and an improved covering power.

闪光/覆盖能力

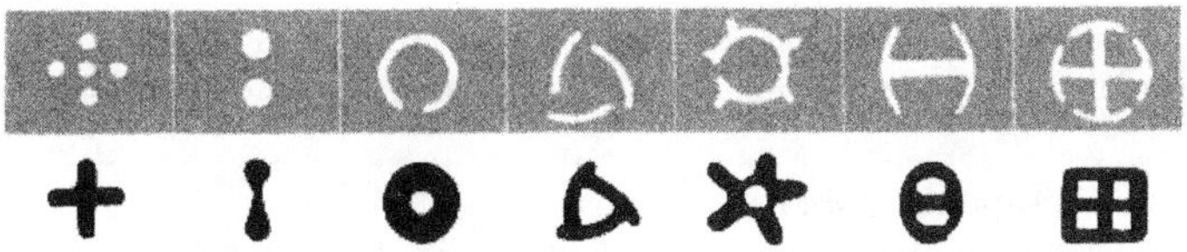

Fig. 2 - 39 Shapes of spinnerettes and fiber cross sections

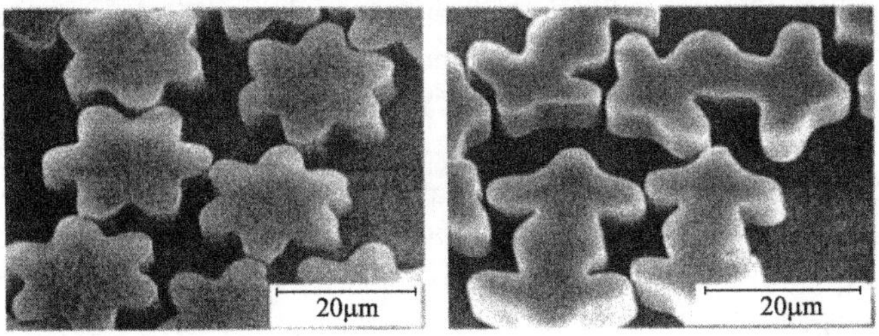

Fig. 2 - 40 Nonround cross section

The main thrust of this development has been in filament fibers, to give a different light deflection off the fiber. Silk has a nonround cross section, roughly triangular in shape, and its beautiful luster sets a standard. When light falls on a rodlike, round cross section fiber and is deflected in broad beams, it gives off a hard shine; with fine filaments of a nonround fiber, light is broken up to give that silk-

柔和的 like subdued luster. In coarse filaments, a nonround cross section results in a glitter or sparkle; this has been used to make "sparkling nylon" pantyhose or glittering organza for holiday season wear.
连裤袜
透明硬纱

手感光滑 Lobal filaments also have a drier, less slippery hand, again more silklike, plus somewhat improved moisture absorbency and /or wicking action, giving more comfort, less static collection, and better soil release.

闪蝶结构纤维 (2) Morpho-structured Fiber

This is an interesting modification from Kuraray of Japan (see Fig. 2 – 41). The fiber combines different kinds of flat profile and multispiral formations, taken from the shape of the wing scales of the Morpho butterfly; the result in fabric is to give changeable color, like the iridescence of a butterfly's wings.
大闪碟

(a) Morpho butterfly (b) Surface of morpho-structured fiber (c) Cross section of morpho-structured fiber

Fig. 2 – 41 The morpho-structured fiber is flat with a multispiral form to imitate the wing scales of the Morpho butterfly of South America for the same visual effect

扁平或带状纤维 (3) Flat Filament or Tapelike Fiber

When coarse, this form produces a glittery look; a traditional fabric has been called "crystal acetate". Rayon with a wide, flat cross section can give a linen-like look and hand to a fabric. Very wide, flat rayon fiber is used for straw-like hats in millinery. Flat rayon filaments are also used extensively to replace the wiry hair fibers, mohair and horsehair, in much tailor's "hair" canvas interfacing; these are cheaper, but cannot be molded with steam in tailoring as the hair fibers can.
女帽类

含空隙的化学纤维 (4) Air Spaces Incorporated into Manufactured Fibers

There are a variety of ways to introduce air spaces into a fiber for a range of characteristics.

中空纤维 ① Voided Fibers

Voided fibers, with hollow channels or air spaces inside, have been used for carpets and other furnishings, since they have soil-hiding ability (see Fig. 2 – 42). They may also be made antistatic and/or soil resistant.

绝热用中空纤维 ② Hollow Fibers for Insulation

These are widely used for their air-trapping ability and lightness, as fiberfill for pillows, comforters, mattress pads, and quilted outerwear. The fiber may have a hollow center, rather like macaroni, or may have four or more channels running
被/床垫/夹层外衣

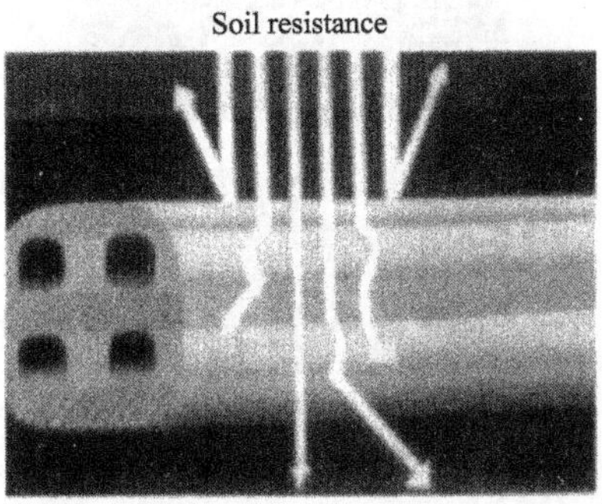

Fig. 2 - 42 Voided fiber hides soil

through it (see Fig. 2 - 43). The extra channels add insulation for cold weather clothing. They also give better loft and resistance to crushing, and so are suitable especially for pillows and comforters.

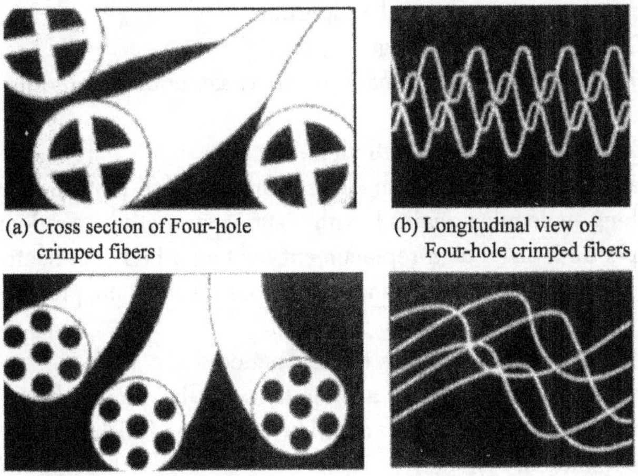

(a) Cross section of Four-hole crimped fibers
(b) Longitudinal view of Four-hole crimped fibers
(c) Cross section of Seven-hole crimped fibers
(d) Longitudinal view of Seven-hole crimped fibers

Fig. 2 - 43 Four-hole crimped fibers and seven-hole crimped fibers

③ Crater Fibers 坑穴纤维

These have a pitted surface, which increases absorbency. A specialized fiber shown in Fig. 2 - 44 is both cratered and hollow.

有凹痕的
坑穴的

(5) Microfibers and Very Fine Fibers 微细纤维和超细纤维

A microfiber is taken to be one finer than silk, approximately 1 dtex. Conventional fiber spinning through finer spinnerette holes, plus drawing, can produce filaments down to about 0.3 dtex; ultra-microfibers down to 0.1 dtex (3 μm diame-

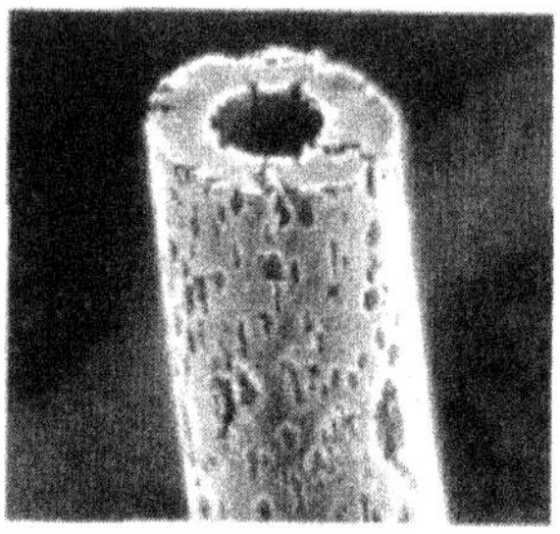

Fig. 2 – 44 Crater fiber

ter) and lower are made by other means, e. g., bicomponent fibers. Heralding ultra-ultra-microfibers are nanofibers, possible by technology that could produce fibers down to 3 nm (0.0000001 dtex).

纳米纤维

2.4.2 Physical or Chemical Properties Modified

物理或化学性能改性

Fibers are modified by the addition of small amounts of selected chemicals. Performance characteristics that can be changed include dyeing behavior, strength, pilling, texture, bulk, and thermal properties.

高强纤维

(1) High-Tenacity (High-Strength) Fibers

High-tenacity fibers have been made in many generics, including rayon, nylon, and polyester.

高密度聚乙烯

(2) HDPE (High-Density Polyethylene)

The strongest, lightest fiber yet made is a form of high molecular-weight polyethylene (extra-long molecular chains) with "super strength". Trademark names are Spectra®. Spectra is used as a replacement for tough body connectors such as tendons and for protective clothing (shrapnel-proof vests), to protect from shearing forces.

防榴散弹背心/剪切力

抗起球纤维或仿麂皮纤维

(3) Low-Pilling Fibers or Fibers to Be Sueded

These are achieved by providing a fiber that can be broken so that pills can be brushed off in the manner of wool, or allowing easier finishing in giving a suede-like or softened surface.

仿麂皮

易染纤维

(4) More Dyeable Fibers

More dyeable fibers are made in a number of generics, but particularly in polyester. Fiber dyeability is altered by the introduction of chemicals that add sulfonic groups to the molecule and by the substitution of isophthalic acid for a small portion of the terephthalic acid. These changes create fibers that can be dyed by cationic dyes instead of disperse, which is the type of dyestuff most frequently used for polyester.

磺酸基

异酞酸

对苯二甲酸/阳离子染料

Acid-dyeable rayon is made so that it can take up the same type of dye as wool in blends.

抗静电纤维

(5) Antistatic Fibers

Antistatic fibers often contain a very small amount of material such as carbon that will conduct electricity; this is termed an epitropic fiber—one with an altered surface property.

Review Questions

思考题

1. What are textile fibers? How would you demonstrate this with a piece of woven fabric?

2. What are the two main sources of natural fibers?

3. What is the first manufactured fiber? What is the first synthetic fiber? What is the difference between them?

4. What is the basic unit of the long-chain cellulose medecule? What other familiar material is made up of the same type of unit?

5. What is the general name for fibers taken from the stems of plants?

6. What is the unit of a protein called?

7. What is the spinning method for both nylon and polyester?

8. Nylon was the dominant synthetic fiber for a long time; Polyester is by far the most-used. What are its main advantages over nylon?

9. What are the distinctive properties of acrylic fibers?

10. Why is spandex being used in active sports wear such as aerobic exercise costumes?

Suggested Activities

实训题

1. What five fibers are present in most of your own clothing? Make up a table showing the article, the fiber content, and the properties of that fiber that make it suitable for that use.

2. If you have access to a microscope and simple supplies, follow directions for untwisting yarns, and mount fibers in water on slides, with a cover slip over each, to give you a longitudinal view. Compare the principal natural fibers with each other, and with viscose rayon, acetate, nylon, polyester, and acrylic. Compare, if possible, mercerized with untreated cotton. Try to find the "Typical Features" shown in this section.

Chapter 3 Textile Yarns 纺织纱线

To weave or knit a fabric it is necessary to have yarns. Thus, the making of yarn is nearly as old as the manufacture of fabric and definitely predates recorded history. In prehistoric times, fibers were twisted together (捻合) in simple ways to form yarns. It is very likely that these early yarn-making technologies involved merely rolling fibers together between the palms of hands, between the fingers, or between a hand and another part of the body, such as the thigh. As a matter of fact, such techniques are used today by craftsmen and artist-weavers to make special yarns for hand-crafted (手工艺的) fabrics.

It was necessary to invent simple methods of disentangling, separating, and arranging the fibers according to their length, other than by just using the fingers. Thus, crude methods of carding were invented to separate the fibers according to their length of staple. Eventually, techniques were refined. In time, long filament strands unwound from silk cocoons, and still later, filaments formed by chemical synthesis were made into yarns. Now yarns are also made by integrating the staple and filament fibers.

Yarns can be made of staple (short) fibers by any one of several techniques. The method used is dependent upon such factors as the manufacturer's preference of equipment, the economic implications, the fibers to be used, and desired properties of the yarn to be produced.

Ring spinning (环锭纺纱) is the oldest and most widespread technique. Open-end spinning (自由端纺纱) is another major method. Newer methods have been developed, and experiments to find more effective and suitable processes continue.

The following sections are concerned with the processes by which yarns are manufactured and with the yarn variations available. The type of yarn used may produce variation in fabric appearance, durability, maintenance, and comfort. It is important to note also that fiber properties have considerable influence on yarn properties.

3.1 Yarn Processing 纱线加工

3.1.1 Conventional Ring Spinning 传统环锭纺纱

The value and character of a yarn are determined by three factors belowed:

① Kind and quality of fibers.

② Amount of processing necessary to produce fineness.

③ Amount of twist, which increases tensile strength in the finished yarn. 成品纱

The purpose of the yarn must be anticipated, as this determines the number and kind of manufacturing operations.

The formation of yarn from staple fiber by spinning becomes possible when they have surfaces capable of cohesiveness. This quality is exemplified by the natural twist of the cotton fibers which enables them to entwine around each other, the roughness of the linen fibers which cause them to cling together, and the scales of the surfaces of the wool fibers which cause them to grasp each other. Flexibility permits the fibers to be twisted around one another. Uniformity of staple gives yarns a required evenness and improves the quality.

Because the production of cotton yarn lends itself to a simple description of the manufacturing operations that make staple fiber into yarn, cotton fiber has been used in this section to illustrate the spinning process. The schematic of conventional ring spinning is shown in Fig. 3 - 1 to Fig. 3 - 8. It must be remembered, however, that all other fibers pass through spinning operations, although, of course, there are differences that will be explained in the sections dealing with the various fibers. One such difference occurs when long strands, or filament, such as silk, are used rather than staple. In such instances, the required number of filaments is simply twisted together in a ropelike fashion.

The development of short fibers, or staple, into yarn, when stated in terms of basic manufacturing processes, is as follows: blending, opening, and cleaning, carding, combing, drafting, twisting, and winding. As the fibers pass through these processes, they are successively formed into lap, sliver, roving, and finally yarn. Here are the manufacturing operations in which these stages occur:

① Lap to card sliver by the carding process. 棉卷/生条/梳理工序

② Card sliver to comb sliver by the combing process (if the fiber is to be combed). 精梳条子/精梳工序

③ Sliver to roving by the drafting, or drawing-out, process. 粗纱/牵伸

④ Roving to yarn by further drafting and twisting process. 细纱

⑤ Yarn reeled on bobbins, spools, or cones by the winding process. 筒管/有边筒管/锥形筒管/络筒工序

3.1.1.1 Blending, Opening, and Cleaning 混棉、开棉、清棉

The cotton arrives at the mill in large bales weighing about 225 kg each. The compressed mass of raw fiber must be removed from the bales, blended, opened, and cleaned. Blending is necessary so as to obtain uniformity of fiber quality; opening is necessary in order to loosen hard lumps of fibers and disentangle them; cleaning is required to remove trash-such as dirt, leaves, burrs and any remaining seeds — in order to prepare the fiber for spinning into yarn. These functions are accomplished through a continuous series of stages. 棉包

As shown in Fig. 3 - 1, cotton is passed from bales 1 onto apron 2. Apron moves cotton to blending apron 3. Blending apron has sharp spikes that raise cotton until part of it is knocked off by roll 4. Some of the cotton stays on apron. The cotton knocked back by roll 4 continues to churn and blend until picked up again by apron. Another roll 5 strips off cotton that was not knocked back by previous roll 输送帘 混棉帘/角钉

4. Cotton falls on conveyor belt 6 and is carried to next process.

Fig. 3 – 1 Schematic of blending feeder

皮棉/喂棉罗拉
开棉锡林/翼式打手
尘笼
筛网/风扇

As shown in Fig. 3 – 2, lint cotton falls on apron 1 and passes between feeder rolls 2 to beater cylinder 3. The rapidly whirling beater blades take off small tufts of cotton, knock out trash, and loosen up the mass. The two screen rolls 4 are made of screen material, and air is sucked out of them by fan 5. This draws the cotton from beater and condenses it on the surface of the screen rolls from which it is taken and passed on by the small rolls 6. Air suction through cotton takes out dirt and trash. Conveyor belt 7 passes cotton to another type of beater 8 (Beaters shown are typical of the many types used). From the beater, the cotton passes to a conveyor and is carried to picker.

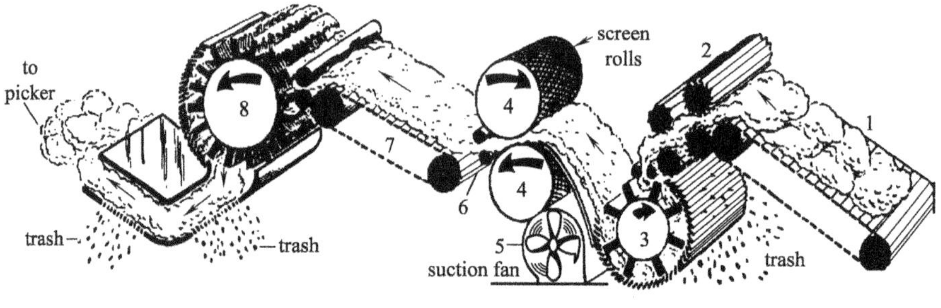

Fig. 3 – 2 Schematic of opener

清棉机

As shown in Fig. 3 – 3, cotton 1 in a loose mass enters picker, which is a series of beaters 2 and screen rolls 3 similar to those described under "opening" but progressively more refined. At the final output of beater and screen system 4, cotton has again been formed into a sheet, or lap. At this point, the "evener" operates to feed more or less cotton to make lap perfectly uniform as it is either wound up into a lap roll 5 on winding rolls 6 and then taken to the carding process, or conveyed as a lap by a chute feeder directly to the card. 卷取辊 管道喂给装置/梳棉机

Fig. 3 – 3 Schematic of picker

3.1.1.2 Carding 梳棉

As shown in Fig. 3 – 4, the lap 1 from picker unrolls, and feed roll 2 passes cotton to the licker-in roll 3 covered with saw-tooth like wire. The licker-in roll passes fibers against cleaner bars 4 and gives them up to large cylinder 5, which passes between the thousands of fine wires on surface of cylinder and on flats 6. The cotton follows large cylinder to doffer cylinder 7, which removes lint from large cylinder, The doffer comb 8 vibrates against doffer cylinder and takes lint off in a filmy web that passes through condenser rolls 9, coiler head 10, and then into can. The sliver may be passed from can to combing for further removal of foreign matter and parallelization of fiber or directly to drawing. 刺辊/锯齿条 除尘刀 梳针/盖板 道夫 道夫斩刀/棉网 轧辊/圈条器/条筒 夹杂物,异物

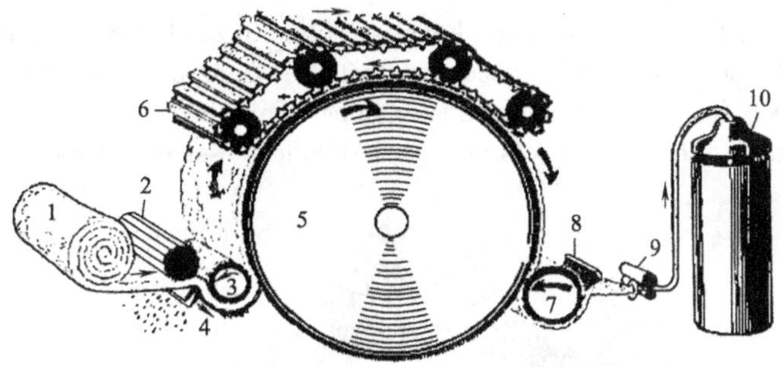

Fig. 3 – 4 Schematic of carding

3.1.1.3 Combing 精梳

When the fibers are intended for fine yarns, the sliver is put through an additional straightening process called combing. In this operation, fine-toothed combs continue straightening the fibers until they are arranged with such a high degree of parallelism that the short fibers, called noils, are combed out and completely separated from the longer fibers. 精梳落棉

The combing process forms a comb sliver made of the longest fibers, which, in turn, produces a smoother and more even yarn (Fig. 3 – 5). This operation eliminates as much as 25 percent of the original card sliver; thus almost one-fourth of the raw cotton becomes waste. The combing process, therefore, is identified with consumers' goods of better quality. Since long-staple yarns produce stronger, smoother, and more service fabrics, quality cotton goods carry labels indicating that they are made from combed yarns or combed cottons.

Fig. 3 – 5 Combed slivers are combined as they leave the combing machine to further increase density and compactness of future yarn

并条 3.1.1.4 Drawing

As shown in Fig. 3 – 6, six cans 1 that are filled at cards feed each drawing from 输入装置/导条匙 delivery. The spoons 2 are connected so that if any one of the six slivers from can should break, the machine automatically stops. This device prevents making uneven yarn later. Each of four sets of rolls 3 runs successively faster than preceding

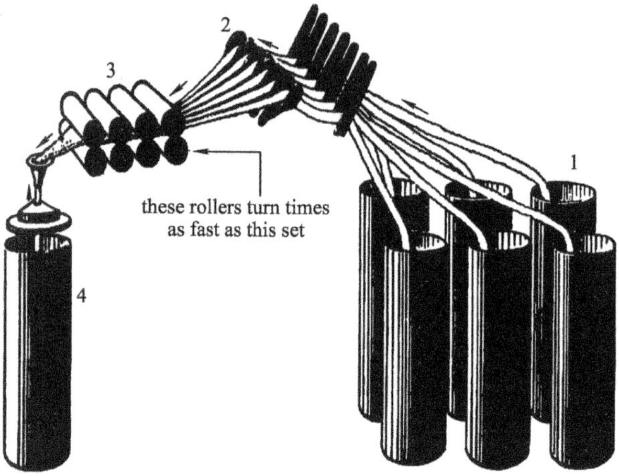

Fig. 3 – 6 Schematic of drawing

set. The last set runs approximately six times as fast as the first set; consequently, sliver coming out is the same size as each one of six going in, but is attenuated to six times the length per minute. The drawn sliver is neatly coiled again in roving can 4 by coiler head. The sliver is now much more uniform and fibers are much more nearly parallel. The sliver is now ready for roving frames. In actual practice, drawing is usually repeated. 熟条

3.1.1.5 Roving 粗纱

As shown in Fig. 3 – 7, the can of sliver 1 from drawing frames is fed between three sets of drafting rolls 2. Each following set of rolls runs faster than preceding set. This pulls sliver and thins it down, making fibers nearly parallel. The spindle 锭子 3 turns flyer 5 锭翼 and is driven at a constant speed. The front rolls (nearest flyer) are set at a speed that gives strand coming out of the rolls a predetermined number of turns of twist 捻回 per inch as it moves along between rolls and flyer. The bobbin 4 is driven by a source separate from gear that drives spindle and flyer. The bobbin is regulated to turn automatically at a speed sufficiently faster than flyer, which causes roving to wind on bobbin at same rate it is delivered by front roll.

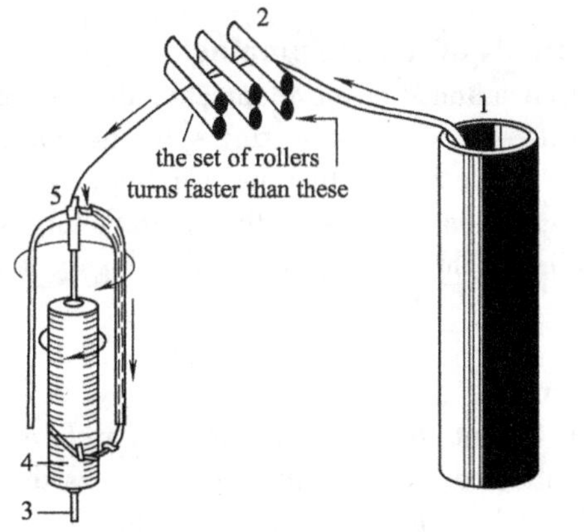

Fig. 3 – 7 Schematic of roving

3.1.1.6 Spinning 细纱

As shown in Fig. 3 – 8, the principle for spinning is the same as that used for roving except that the operation is more refined and a ring and traveler 钢领/钢丝圈 are used instead of the flyer. From bobbin 1 roving is fed between sets of drafting rolls 2 to draw strand down to its final desired size. The spindle 3 turns bobbin 4 at a constant speed. The front set of rolls is adjusted to deliver yarns at a speed sufficient to insert desired amount of twist as strand moves along. The traveler 5 glides freely around ring 6. The tension caused by drag of traveler causes yarn to wind on bobbin at same rate of speed as it is delivered by rolls.

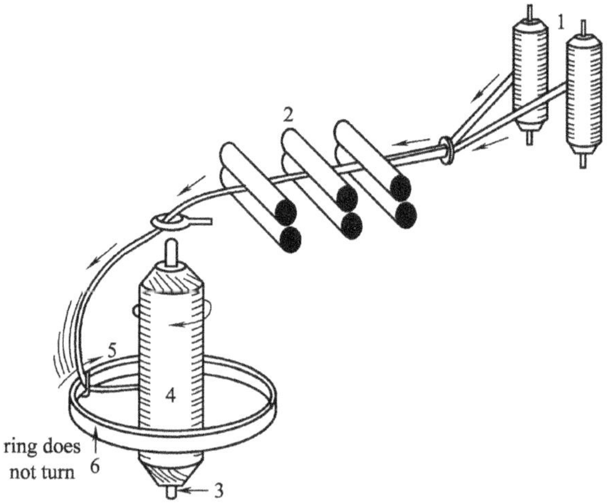

Fig. 3 – 8 Schematic of spinning

新型纺纱方法

3.1.2 Newer Methods of Yarn Spinning

Along with industrial automation development, the ring spinning has protruded difficult to overcome its shorts. Because ring spinning machine possesses intrinsic attribute of an organic whole for twisting and winding, very small package and low spinning speed are its bottleneck to solve the contradictory of package and winding which impede to realize high spinning speed and big package as well automation process.

气流纺纱

3.1.2.1 Rotor Spinning

气流 转杯

A sliver of fibers is fed into the unit, a current of air forces the fibers into a loose form, and they are collected on the rotating device. As they collect they form the yarn, which is pulled off by mechanical means and wound onto yarn packages. Twist is inserted as the yarn is removed from the rotor(Fig. 3 – 9).

卷取筒子

In Fig. 3 – 9 (a), sliver of fibers enters from below and is fed into the rotor, where centrifugal force converts fibers into completed yarns. The take-up package holds yarns ready for further processing into fabric. In Fig. 3 – 9 (b), sliver of fibers enters from above and is fed into the rotating funnel, where centrifugal force converts fibers into completed yarns. Yarns is moved from the base of the funnel.

环锭纺纱

Rotor spinning has become a popular yarn making method. Although there are differences in quality between ring-spun and rotor-spun yarns, both have advantages in various end-uses.

An important aspect of rotor-spun yarns is economic. Yarn production is faster, labor needs are reduced, space needs are reduced, energy requirements are reduced, the operation is cleaner, the system can utilize cotton fibers that would not be usable in ring spinning, there are fewer knots, and less waste occurs. A major disadvantage, as yet, of rotor-spun yarns is that it is difficult to make fine-quality yarns compared to those that can be made by ring spinning. Another problem in rotor spinning is that fiber defects cause more problems than in ring spinning.

Properties of rotor-spun yarns compare favorably with ring-spun yarns. Fiber distribution is more uniform in rotor than ring-spun, and fiber migration is less in rotor than in ring-spun yarns. Because of the way fibers arrange themselves in rotor-spun yarns, dyeing behavior may differ from that of ring spun yarns. It has been stated that deeper colors and differential dye techniques are possible with rotor-spun yarns. 纤维转移

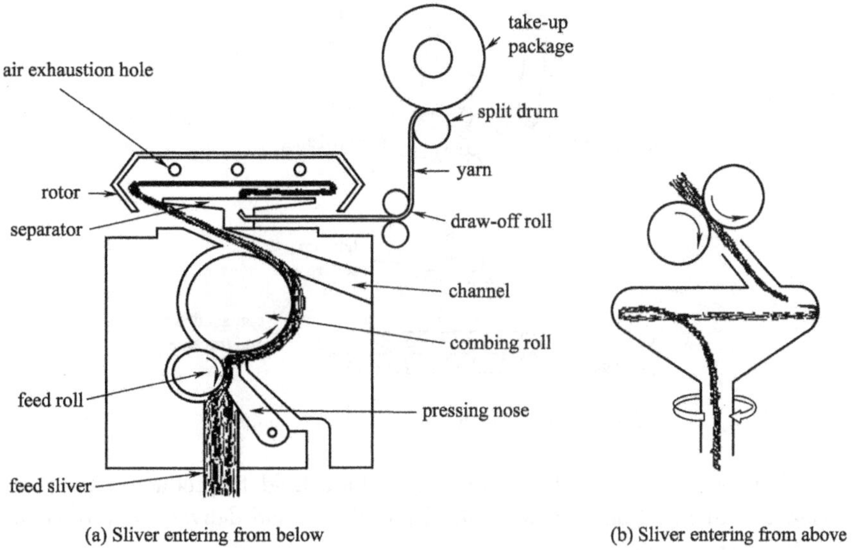

Fig. 3 – 9 Diagrams of two types of units used in rotor spinning

3.1.2.2 Air Votex 涡流纺纱

Spinning of yarns by the air vortex system is very similar to the rotor technique except that the yarn is formed in moving air rather than on a rotating surface (Fig. 3 – 10). The method is not widely used as yet, but it is being promoted as a successful and economic way to produce some core-spun yarns. Yarns made by this method are said to be similar in strength to ring-spun, have greater evenness, better strength and uniformity, and greater bulk. However, these air vortex yarns tend to pill more quickly than other types. 包芯纱 起球

3.1.2.3 Electrostatic Spinning 静电纺纱

The researchers began by exploring the possibility of removing the short cotton fibers from the longer ones by an electrostatic process. The principle is based upon the knowledge that widely differing materials can be separated from each other electrostatically. The idea of separating different sizes of the same material, and more particularly cotton, was a novel approach.

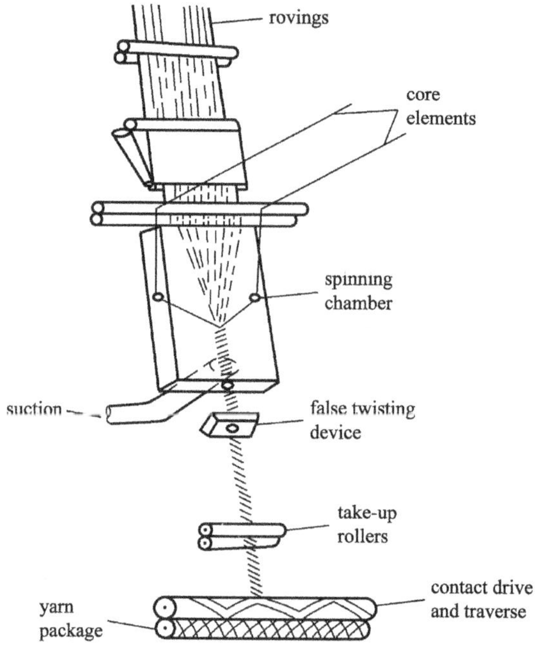

Fig. 3 – 10 The air vortex system of spinning

The electrostatic system utilizes an electrostatic field to produce a parallel arrangement of fibers, transport and collect the fibers, and deliver them to mechanical twisting equipment.

摩擦纺纱

3.1.2.4 Friction Spinning

In the early 1970s Dr. Ernst Fehrer received an Austrian patent for a friction spinning process which was subsequently commercialized under the trademark DREF. The system has since been modified and is identified as DREF II, a variation of the open-end spinning process. The principle of the technique is that of passing along the longitudinal drum axis a stream of carded fibers in the roll nip of the perforated drums, both of which are moving in the same direction. This action forces the fiber to be wedged along the nip, thereby becoming compressed; the friction of the rollers cause the fibers to twist around each other.

尘笼轴向

The process utilizes a current of air to transport the carded fibers to the longitudinal nip region of the two drums. To improve parallelization, parallel orienting disks with fingers are used to align the fibers in the take-off direction of the yarn. The air stream gives the sliver end the initial twist as it presses the disentangled fibers against the continuously forming yarn compound wedged into the roll nip. Air suction through the perforated drums produces a uniform twist effect as the yarn is drawn off the end of the drums. The amount of twist may be controlled by regulating the suction. The yarn count is determined by the thickness of the wedge in the yarn-forming nip and the width (diameter) of the nip itself (Fig. 3 – 11).

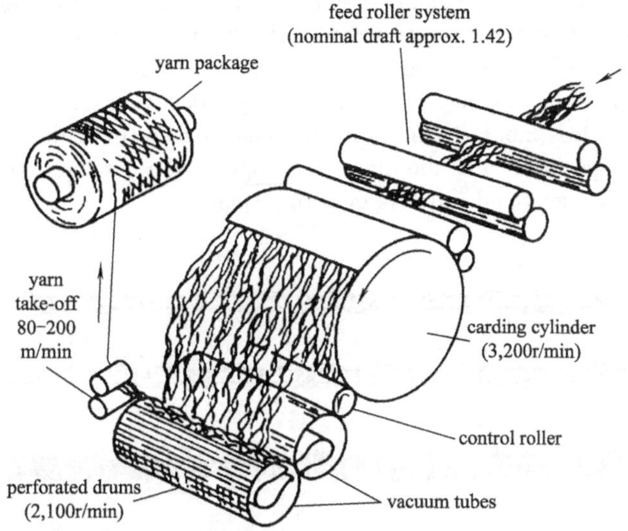

Fig. 3 – 11 Schematic of DREF friction spinning process

As shown in Fig. 3 – 11 fibers are fed from a sliver to a carding cylinder where individual fibers are spread out. As the cylinder turns, the fibers are pulled away and taken to an area where a jet of air blows the fibers off the rolls and onto perforated drums. Suction units hold the fibers to permit twisting and rolling off to form the yarn. 分梳辊 多孔尘笼 抽吸装置

The resultant yarns are bulky and have surface and mechanical properties similar to those of woolen yarns. And since they do not have the wrapper fiber effect that is achieved by the rotor open-end spinning, they have a low tensile strength. Improvements in the technique continue to be developed and there are a number of variations of this system.

Novelty specialty yarns, such as boucle and nub, can be produced. Also, by feeding a filament core yarn axially into the center of the spinning zone, core spun yarns can be produced. 芯纱

In 1979, Fehrer introduced the DREF III, an outgrowth of the DREF II. Yarns are spun from a wide range of fibers including cotton, rayon, nylon, aramid, polyester, acrylic, and polypropylene or blends of fibers into counts ranging from 32 – 97tex. The process contains a dual yarn spinning system which can produce yarns by feeding either from two separate slivers or from one staple sliver feed and one filament feed. In the latter instance, the staple completely covers the filament which serves as the core of the yarn.

3.1.2.5 Self-Twist Spinning 自捻纺纱

Self-twist yarns are essentially two-strand (or more) yarns, They may involve

one strand of staple fibers and one of filament, two strands of staple fibers, or two
夹长丝自捻纱 of filament, or multiple combinations of these. A common type is the Selfil yarn, which is composed of one strand of staple fibers with some twist and a filament strand wrapped around the staple strand. The filament strand alternates S and Z twist. A number of combinations of these strands have been made (Fig. 3 – 12).

Self-twist spinning produces yarns more cheaply than ring spinning and considerably faster with less labor. Selfil yarns are slightly more even than ring-spun and 包覆 are much stronger because of the filament wrapping. Furthermore, in knitted structures Selfil yarns snag and pill less than ring-spun.

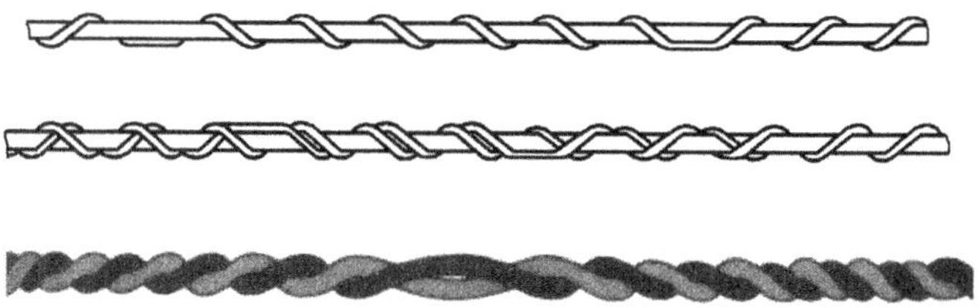

Fig. 3 – 12 Diagram showing examples of self-twist yarns

无捻纺纱

3.1.2.6 Twistless Yarn Spinning

无捻纱/黏合剂 须条 淀粉

In the manufacture of twistless yarns the fibers are held together by adhesive. A roving is made and then drawn to a fine strand; during the drawing an adhesive is applied by rollers. A common adhesive is starch. The drawn yarn coated with adhesive is wound onto packages, the package is steamed, and the adhesive bonds the fibers together.

包芯纺纱

3.1.2.7 Core Spinning

This process is diagrammed in Fig. 3 – 13. It may also be used for a variety of other fibers. Core spun yarns may be made by the air vortex open-end spinning technique, in which case both the sheath and core can be of staple fibers. A process called coverspun® has been used to create a yarn with a wool core wrapped with filament fibers. The wrapping or sheath may be of any man-made filament fiber.

纺纱的进展

3.1.2.8 Development of Spinning

The conventional ring spinning technique is currently the most widely used, accounting for an estimated 90% of staple yarn market. The remaining systems are often referred to as newer methods stated in the above sections.

To improve yarn properties, there are some modification to the conventional ring spinning such as Sirospun, Sirofil, Solo spinning, and Compact spinning. Their advantages are shown in Table 3 – 1.

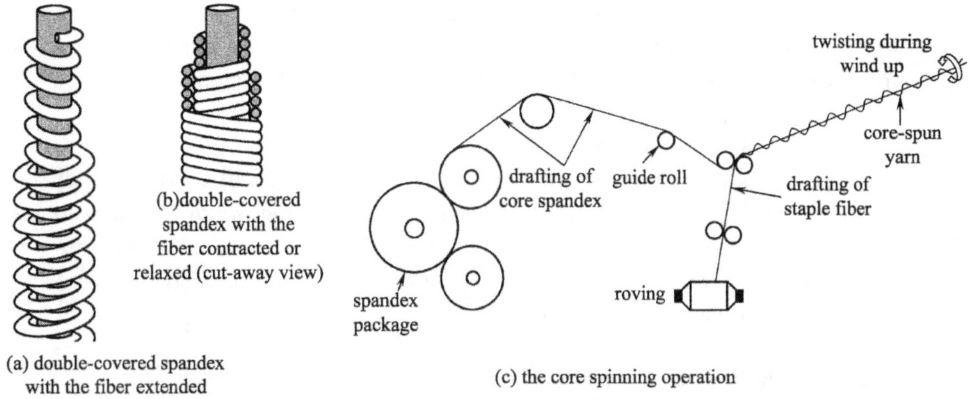

Fig. 3 – 13 Diagrams showing core-spun yarns and core spinning process

Table 3 – 1 Advantages of Sirospun, Sirofil, Solo Spinning and Compact Spinning

Modification	Advantages	
Sirospun	the ring spinning frame can be easily modified; yarn structure is similar to plied yarn; increasing wear-resisting and decreasing the hairs	赛络纺纱
Sirofil	enhancing strength of composite yarn; decreasing yarn cost; improving yarn evenness	赛络菲尔纺纱
Solo spinning	higher spinning productivity and efficiency; elimination of twisting and associated process; higher winding productivity and efficiency; finer yarn at lower cost	索罗纺纱
Compact spinning	decreasing yarn hairiness; increasing yarn strength and elongation; improving yarn evenness	紧密纺纱

In the open-end spinning methods, only two have achieved commercial success: rotor spinning and friction spinning. Of the two, rotor spinning is the more widely used commercially, because a wider range of yarn counts can be spun with unique and suitable yarn properties. Compared with ring spinning process, new spinning methods realize high spinning speed, big package and automation as well, but show their limitation in fine yarn range. Efforts should be made to develop the yarn of chemical fibers and to expand the yarn for industrial and decorative use for rotor spinning; to increase yarn count range, to improve the twist efficiency and to reduce noise and energy consumption for friction spinning.

蚕丝纱加工 3.1.3 Silk Filament Yarn Processing

The cocoons that are raised by silk farmers are delivered to a factory, called a
缫丝厂 filature, where the silk is unwound from the cocoons and the strands are collected
绞丝 into skeins. Some cocoons are scientifically bred in such factories.

选茧 3.1.3.1 Sorting Cocoons

The cocoons are sorted according to color, size, shape and texture, as all these affect the final quality of the silk. Cocoons may range from white or yellow to grayish, depending on the source and the type of food consumed during the worm stage. Cocoons from China are white; Japanese cocoons are creamy white and yellow; Italian cocoons are yellow.

丝胶软化(或煮茧) 3.1.3.2 Softening the Sericin

After the cocoons have been sorted, they are put through a series of hot and cold
浸渍 immersions, as the sericin must be softened to permit the unwinding of the filament as one continuous thread. Raw silk consists of about 80 percent fibroin and 20 percent sericin. At this time, only about 1 percent of the sericin is removed, because this silk gum is a needed protection during the further handling of the delicate filament.

缫丝 3.1.3.3 Reeling the Filament

The process of unwinding the filament from the cocoon is called reeling. The
疵点/生丝 care and skill used in the reeling operation prevent defects in the raw silk. As the filament of a single cocoon is too fine for commercial use, three to ten strands are usually reeled at a time to produce the desired diameter of raw silk thread. The cocoons float in water, bobbing up and down, as the filaments are drawn upward
瓷眼,集绪器 through porcelain eyelets and are rapidly wound on wheels or drums while the operator watches to detect flaws. As the reeling of the filament from each cocoon nears completion, operators attach a new filament to the moving thread. Skilled operators have an uncanny ability to blend the filaments, while always retaining the same diameter of the rapidly moving silk strand. The sericin acts as an adhesive. It aids in holding the several filaments together while they are combined to form the single thread. On old-style reeling machines at high speed, an operator could handle five to seven threads, and on the newest models, twenty-five threads.

The usable length of the reeled filament is from 300 – 600m (1,000 – 2,000 feet) approximately a quarter of a mile long. The remaining part of the filaments
绢丝 used as valuable raw material for the manufacture of spun silk.

缫制的丝 The term "reeled silk" is applied to the raw silk strand that is formed by combining several filaments from separate cocoons. The diameter of the silk fiber is so fine that an estimated 3,000 cocoons are required to make a yard (1 m) of silk fabric. The silk filaments are reeled into skeins, which are packed in small bun-
小包丝 dles called books, weighing 2.0 – 4.5kg (5 – 10 pounds). These books are put
大包丝,件丝 into bales, weighing about 60 kg (133.33 pounds). In this form, the raw silk is shipped to all parts of the world (Fig. 3 – 14). From the filature, the books of

reeled silk are put through the following manufacturing processes.

Fig. 3 – 14 Reeling silk cocoons

3.1.3.4 Degumming of Thrown Silk

加捻丝的脱胶

Thrown silk yarns still contain some sericin that most be removed in another soap bath to bring out the natural luster and the soft feel of the silk. As much as 25 percent of the weight is lost by the degumming process. When the gum has been removed, the silk fiber or fabric is a creamy white color, beautifully lustrous, and luxuriantly soft.

Degumming may take place after the silk yarn is thrown in order to prepare it for yarn dyeing, or it may be done in any finishing process after the fabric is woven. For example, the tightly twisted yarns used for crepe effects still contain sericin. A small amount of sericin is sometimes left in the yarn or in the fabric to give the finished product added strength or a dull finish. Ecru and souple are examples of silk fabrics from which sericin has been partially removed.

消光整理/本色整理/半练整理

3.2 Yarn Construction and Properties

纱线结构与性能

3.2.1 Yarn Twist

纱线捻度

As fibers, staple or filament, are formed into yarns, twist is added to hold the fibers together. The amount of twist is sometimes suggested broadly by such terms as low, medium, and high, but it is more accurately indicated by the number of turns per inch. The turns per inch (TPI) needed to form the best possible yarn varies with the yarn diameter. As the yarn becomes finer, it requires more twist. To determine the optimum degree of twist, a factor called the twist multiplier(TM) is used. The TM ranges from about 3 (for soft yarns) to 6 or more (for hard-twist

捻/英寸

捻系数

弱捻纱/强捻纱

yarns). We can use the twist factor either to calculate the TPI required to give the desired degree of twist in any counts, or to compare the degree of twist in yarns of different counts.

The strength of yarns is due, in part, to the amount of twist that has been imparted. Strong yarns require considerable twist. However, beyond an optimum point additional twist will cause yarns to kink and finally to lose strength.

The twist factor which gives the maximum strength in any given staple yarn is sometimes called the optimum twist factor.

最佳捻系数,临界捻系数/左手捻(S捻) 右手捻(Z捻)

The direction of twist is also important. Yarns can be twisted with either a left-hand twist (S twist) or a right-hand twist (Z twist). The direction of the twist conforms to the center bar of the letter (Fig. 3 – 15).

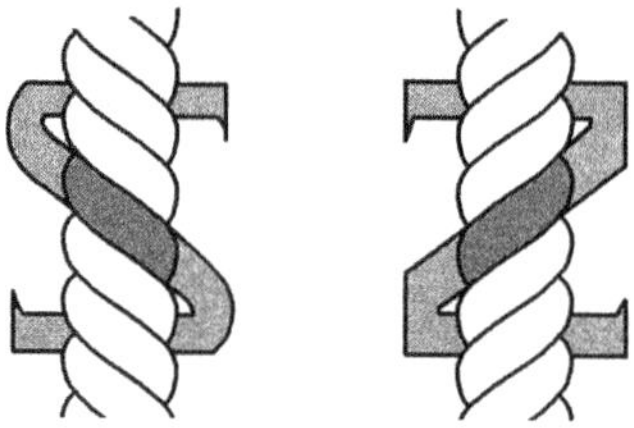

Fig. 3 – 15 S and Z twist in yarns

Various effects can be obtained by combining yarns of different twist direction, and durability may be increased by efficient plying of S and Z twist single yarns.

纱线细度

3.2.2 Yarn Number

Yarns are bought and sold by weight, but the weaver is vitally interested in the length of yarn he will obtain in a given weight, because the amount of fabric which he can make depends primarily on the length of yarn available for making it. Therefore, it is natural to describe the fineness or coarseness of yarns in terms of weight per unit length, or of length per unit weight. The fineness of a yarn expressed in one of these ways is its yarn number or count.

单位长度的重量/单位重量长度/纱线支数

Very many different systems of yarn numbering have been evolved in the course of time in the different localities where the textile industry has become established. They can all be classified under the headings of direct (based on weight per unit length) or indirect systems (based on length per unit weight). Two important examples of each are given below.

直接制(定长制)

3.2.2.1 Direct Systems

旦尼尔制

(1) The denier system

The denier of a fiber, filament or yarn is defined as the weight in g of 9,000 m or, alternatively, the weight in mg of 9 m.

特克斯制

(2) The tex system

This is the proposed universal system which is intended eventually to replace all other systems. The tex number of a yarn is the weight in g of 1,000 m, i. e., g/

km. Clearly, there is a simple relation between denier and tex, namely denier = 9tex. For example, the denier equivalent of 180 tex is 20 denier. In other words, 180 tex and 20 denier are alternative ways of stating the count of a particular yarn.

In any direct system, the weight per unit length is directly proportional to the yarn number. A higher yarn number implies a coarser yarn. Yarn numbers expressed in a direct system are additive, i. e. , the number of a composite yarn is the sum of the numbers of its component yarns. For example, the denier of a two-fold yarn composed of one 75 denier yarn and one 150 denier yarn is 225 denier, neglecting any take-up in twisting. Further, if the take-up or contraction in twisting were, for example, 10 percent, the actual denier of the two-fold yarn would be 225 plus 10 percent, or 247.5 denier. Calculations in a direct system such as denier or tex are simple.

3.2.2.2 Indirect Systems

间接制(定重制)

(1) The cotton system

棉英制

Cotton count is defined as the number of hanks, each 840 yd. long, which weigh 1 lb. For example, a 30 S cotton count yarn has 30 × 840 = 25,200 yd./lb.

绞

(2) The worsted system

精纺毛英制

Worsted count is defined as the number of hanks, each 560 yd. long, which weigh 1 lb. For example, a 30 S worsted count yarn has 30 × 560 = 16,800 yd./lb.

In any indirect system, the yarn number is indirectly proportional to the weight per unit length of the yarn, and a higher number implies a finer yarn. Indirect yarn numbers are not additive. To find the resultant count of a folded yarn, one must add the reciprocals of the component counts and take the reciprocal of the sum. For example, the resultant count of a two-fold cotton yarn composed of one yarn of 20 S and one of 30 S is 12 S, because $\frac{1}{20}+\frac{1}{30}=\frac{1}{12}$.

3.3 Simple yarns

普通纱线

Yarns that are even in size, have an equal number of turns per inch throughout, and are relatively smooth are called simple yarns. A simple, single yarn is the most basic assemblage of fibers suitable for operations such as weaving and knitting.

集合体

A simple, ply yarn is composed of two or more simple single yarns plied or twisted together. In naming a ply yarn the number of singles included usually precedes the word ply. For example, if two singles are used, the yarn is called a 2 – ply; if four are used, a 4 – ply(Fig. 3 – 16 and Fig. 3 – 17).

合股线

双股

四股

Simple cord or cable yarns are composed of two or more ply yarns twisted together(Fig. 3 – 18 and Fig. 3 – 19). In identifying a cord it is necessary to indicate the number of plies and singles in the cord. Thus, a "3, 5 – ply cord" indicates that each ply has five singles and that three of these 5 – ply yarns are combined in making the cord.

绳线

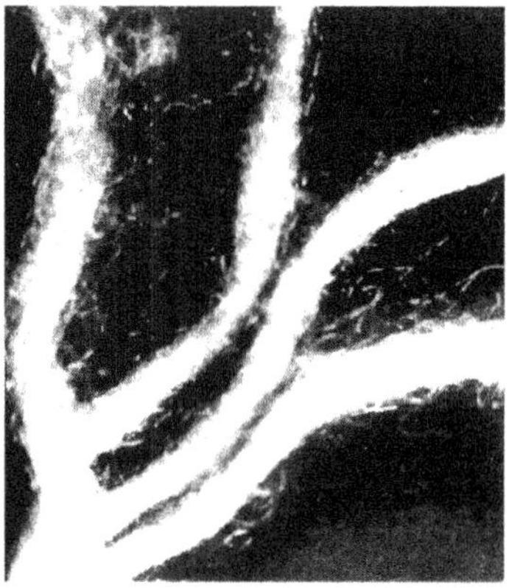

Fig. 3 – 16 Simple yarn, magnified. A 4 – ply yarn with the singles separated; fibers are slightly separated in one of the singles

Fig. 3 – 17 Combing two singles to form a ply yarn. The two yarns are twisted together as they are wound around the bobbin

Fig. 3 – 18 A 3, 2 – ply cord yarn; three plies, each made of two single yarns

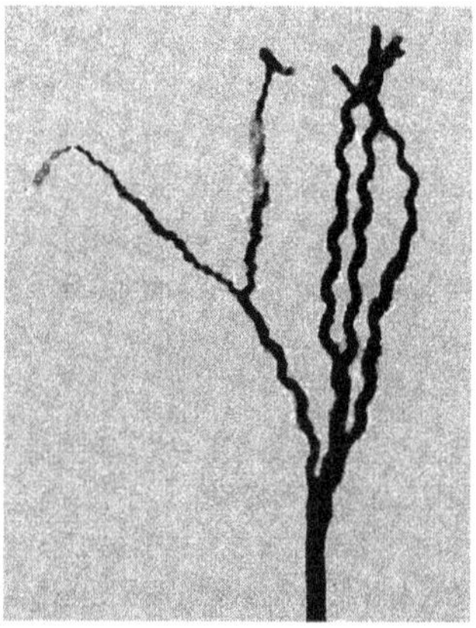

Fig. 3 – 19 A 4, 2 – ply cord yarn; four plies, each made of two single yarns

绉线 扭结 绉织物 花式纱线

Crepe yarns are a variation of simple yarns. However, a crepe yarn possesses a high degree of twist, so the yarn tends to kink. This kinkiness results in the rough texture characteristic of crepe fabrics. Nevertheless, the yarns are evenly twisted and even in size, so they are truly simple yarns rather than complex or novelty yarns.

变形效应

Except for the textured effect achieved by crepe yarns, simple yarns in themselves do not create variation in fabric appearance. However, a combination of simple yarns of different size, different amounts of twist, and/or different fiber content can produce many interesting effects. The arrangement of yarns in groups (as in dimity) can also yield visual variety. Other changes in appearance depend on the number of warp and filling yarns per inch and the type and amount of twist. Generally speaking, simple yarns produce smooth fabrics. Again, the arrangement of the yarns and combinations of various sizes in the fabric construction will influence the end product. 麻纱,条格麻纱 终端产品

Simple yarns are usually considered to be durable, although the durability is affected by such factors as yarn number, amount of twist, and whether the yarn is single, a ply, or a cord. However, the uniformity of the yarn helps prevent snagging and tearing. Except for the highly twisted crepe yarns, which tend to shrink during care, simple yarns are most often easy to maintain. The arrangement of the yarns, the ply or cord construction, and the degree of yarn balance will dictate maintenance procedures to some degree. 擦毛 撕断/收缩

The statements made concerning simple yarns are generalizations only and may be incalculably affected by differences in fibers, fabric structure, coloring methods, and finishing processes. It is, therefore, essential that the student of textiles be aware of the influence of other fabric "dimensions" on durability, maintenance, appearance, and comfort before establishing specific statements relating to any fabric.

3.4 Complex Yarns 花式纱线

Complex or novelty yarns are usually uneven in size, varied in color, or modified in appearance by the presence of irregularities deliberately produced during their formation. Single novelty yarns are produced by the inclusion of slubs or knots. In plied complex yarns the irregularities are the result of variable delivery of the component parts and irregularities in size of the components. 粗节 结子

Complex yarns are made for their appearance value. These yarns may be single, plied, cord, or modified cord construction. Complex ply yarns are usually composed of the following: a base or core, an effect, a tie or binder yarn (Fig. 3 - 20). 基纱或芯纱/饰纱/固纱

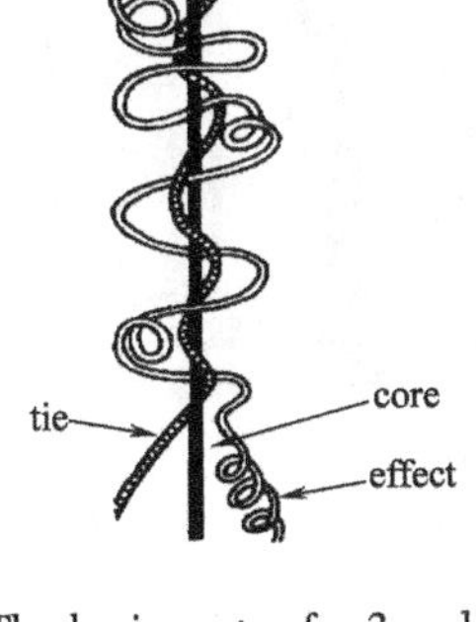

Fig. 3 - 20 The basic parts of a 3 - ply complex yarn

The base yarn controls the length and stability of the end product. The effect or fancy yarn forms the design by the way in which it is applied to the base yarn. The tie or binder yarn is used to attach the effect yarn so that it will remain in position during use and care. The strength of all complex yarns is variable.

应力分布 Irregularities in amounts of twist and size produce an uneven stress distribution, which may result in yarns with some areas considerably weaker than others.

There are several systems used in classifying and naming complex yarns, and definitions of the various types differ.

花式单纱 3.4.1 Complex Single Yarns

竹节纱/竹节丝 植绒纱 Complex single yarns include slub yarns[Fig. 3 – 21(a)], thick-and-thin yarns and flock yarns.

花式股线 3.4.2 Complex Ply Yarns

Complex ply yarns include following yarns:

圈圈线 ① Boucle yarns.
花式圈线和粗松圈线 ② Ratine and gimp yarns.
小毛圈线 ③ Loop or curl yarns[Fig. 3 – 21(b)].
疙瘩线和结子线 ④ Nub(or spot) and knot (or knop) yarns[Fig. 3 – 21(c)].
混色花线 ⑤ Grandrelle yarn.
螺旋花线 ⑥ Spiral or corkscrew yarns.
雪尼尔线 ⑦ Chenille yarns[Fig. 3 – 21(d)].
粗节花线和黑白竹节纱 ⑧ Seed and splash yarns.

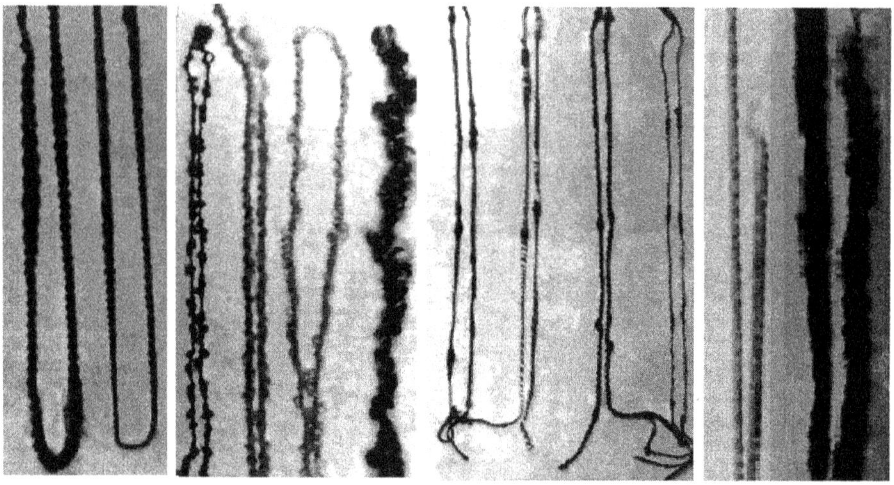

(a) Complex slub yarns　(b) Loop type complex ply yarns　(c) Nub or knot complex ply yarns　(d) Chenille type complex yarns

Fig. 3 – 21 Complex yarns

3.5 Silk Filament Yarns

蚕丝纱

3.5.1 Thrown Silk

加捻丝

Reeled silk is transformed into silk yarn—also called silk thread—by a process known as throwing. The term is derived from the Anglo-Saxon word "thrawn" meaning "to twist". Persons engaged in this work are called throwsters. Silk throwing is analogous to the spinning process that manufactures cotton, linen, or wool fibers into yarns. Unlike those fibers, the manufacture of silk yarn does not include the processes for producing a continuous yarn by carding, combing, and drawing out. The raw silk skeins are sorted according to size, color, and length or quantity, then soaked in warm water with soap or oil. This softening of the sericin aids in handling the thread. After mechanical drying, the skeins are placed on light reels from which the silk is wound on bobbins.

打线,捻丝
捻丝工

During this winding operation, single strands may be given any desired amount of twist. If two or more yarns are to be doubled, they are twisted again in the same direction or in a reverse direction, depending on the kind of thread to be made. To equalize the diameter, the yarn is run through rollers. The thread is then inspected and packaged and is ready for shipment to manufacturers for construction into fabrics.

3.5.1.1 Kinds of Thrown Silk Yarns

捻线类型

The type of yarns and the amount of twist depend on what the weaver desires. Several kinds of silk yarns, or threads, which are used in the manufacture of silk goods are shown in Fig 3 – 22 and listed below:

① Singles[Fig. 3 – 22(a)]. 单根生丝
② Tram[Fig. 3 – 22(b)]. 并捻线
③ Voile[Fig. 3 – 22(c)]. 巴里线
④ Georgette or Crepe de Chine[Fig. 3 – 22(d)]. 乔其线,双绉线
⑤ Organzine[Fig. 3 – 22(e)]. 双股捻线
⑥ Grenadine[Fig. 3 – 22(f)]. 多股捻线
⑦ Two by Two Grenadine Type[Fig. 3 – 22(g)]. 2×2 加捻线
⑧ Compensene S[Fig. 3 – 22(h)]. S 捻平衡线
⑨ Crepe[Fig. 3 – 22(i)]. 绉线

3.5.1.2 Count of Thrown Silk

捻线细度

The size or the yarn count of silk threads is based on a system of weight known as denier stated in section of 3.2.2.1. A denier represents a weight of 0.05 g and the standard length of a raw silk skein is 450 m. An increase in weight of a standard length indicates an increase in diameter and, therefore, denier. A 450 – meter skein weighing 0.05 g would be of a thread of one denier, a skein of 0.50 g would be of a thread of 10 denier, and so forth. Another way of expressing the weight is in g per 9,000 m. And since the tex system measures in g per 1,000 m and decitex (dtex) in g per 10,000 m, conversion from denier to decitex can be done by

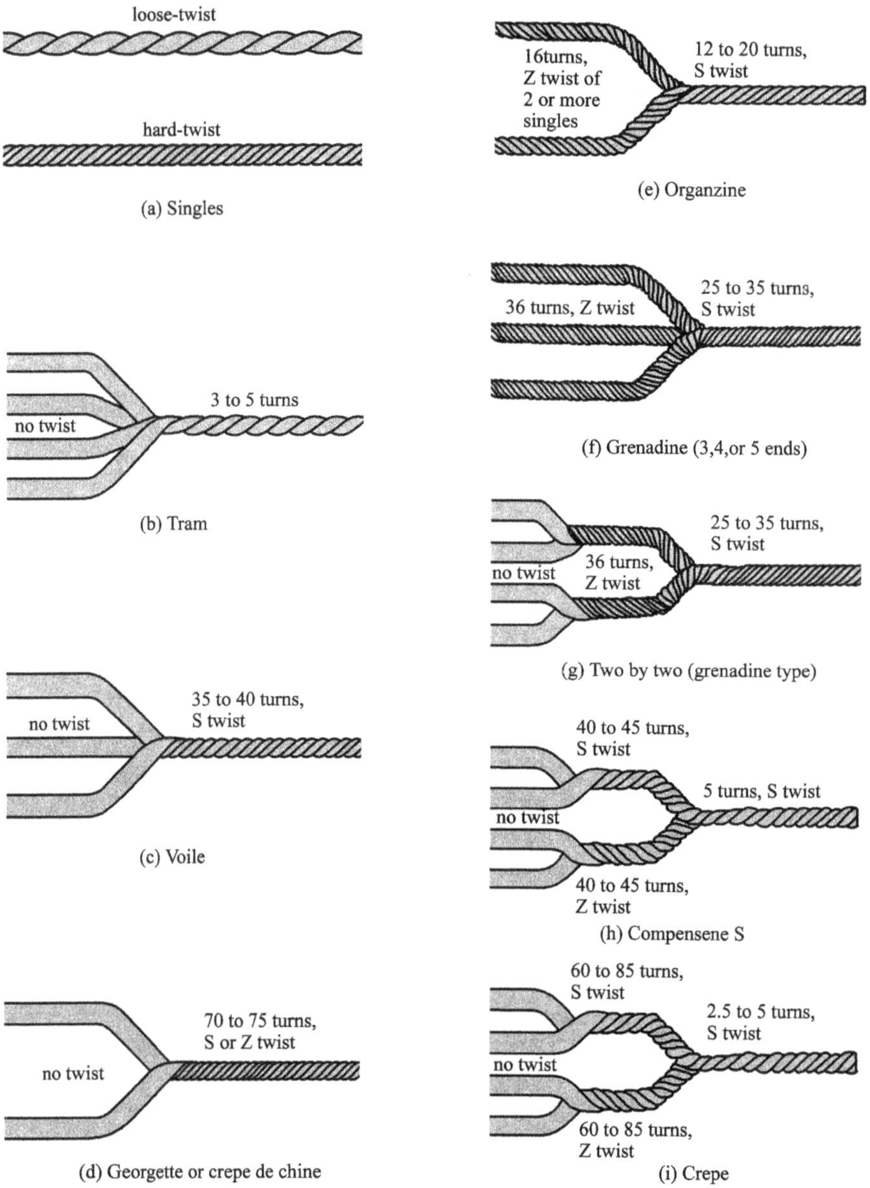

Fig. 3 – 22 Types of thrown silk yarns. "Turns" means turns per inch

multiplying the denier by a factor of 1.111.

The most common size of raw silk is 21 denier, which is expressed as 20 – 22 denier (22.2 – 24.4 dtex) to allow for variation in the diameter of the yarn. The greater the number of cocoons used, the thicker the yarn and the higher the denier (dtex).

3.5.2 Spun Silk

绢丝

Short lengths of inferior silk filaments obtained from waste material are not used in producing reeled silk. After the short lengths have been carded and combed, they are spun together much as cotton, linen, or wool yarns are spun. Spun silk yarns are soft, but they are less lustrous than reeled silk and are not as strong or elastic. Spun silk fabric tends to become fuzzy after wearing because the yarn is made of short staple.

There are several sources of staple silk:

(1) Pierced cocoons, the result of breeding moths that have emerged from their cocoons. 蛾口茧

(2) Double cocoons, the result of two cocoons having been spun by two silkworms too close together (sometimes called douppioni silk). 双宫茧 双宫丝

(3) Floss, brushed from cocoons before reeling. 绪丝

(4) Frison, the coarse and uneven silk fiber at the beginning and end of each cocoon. 茧衣

(5) Scrap, the machine waste left from reeling, throwing, and the like. 碎料

Spun silk is less expensive than reeled silk. Although spun silk has less strength and elasticity because of the shorter staple used, it possesses all the general characteristics of reeled silk. Spun silk is used for shantung and pile fabrics; for dress trimmings, linings, elastic webbing, and sewing silk; for wash silks, velvets, umbrella fabrics, and insulative materials. 起绒织物/外衣装饰/弹性织物/耐洗绸缎/丝绒/伞绸

The fibers are processed into yarns on the principles used in the spinning of cotton yarn: picking, combing, dressing, separation into a lap form, drawing out into sliver form, condensing, drawing, drafting, attenuating, doubling, redoubling, roving, and finally spinning into yarn.

The waste derived from the processing of spun silk yarn is also used. The Trade Practice Rules of the FTC require that such fiber be labeled as waste silk, silk waste, noil silk, or silk noil. For obvious reasons, the last two designations are most frequently used. Silk noil may be reprocessed into spun yarn and woven into textured fabrics for draperies, upholstery, and sportswear. These fabrics are dull, rough, have a cottonlike appearance, but are generally more resilient. 联邦贸易委员会/废丝/废丝/紬丝/紬丝 挂帷织物/运动服

The method of determining the size of spun silk yarn is similar to that used for cotton yarn. In cotton, the term "60 S/2" signifies a 2 – ply yarn consisting of two single strands twisted together, each having a yarn count of 60 S (9.8 tex), resulting in the equivalent thickness of a 30 S (19.7 tex).

In spun silk, a 2 – ply yarn is indicated by the figures "60 S/2". Although this term appears to be the same as the cotton term, it does not mean the same. In spun silk, 60 S/2 means that two yarns, each of a yarn count of 120 S (4.9 tex), have been twisted together to produce a ply yarn that has a count of 60 S (9.8 tex).

3.5.3 Doupion

双宫丝

This silk thread is reeled from two cocoons that have nested together. In spinning, the double thread is not separated. The yarn is uneven, irregular, and the diameter is large. It is used in cloth of this name. The fabrics are very irregular in surface effect. Doupion is popular in suiting, used either alone or in combination with wool, worsted, Tussah, and man-made fibers in fabric blends.

柞蚕丝 3.5.4 Tussah or Wild Silk

Tussah is as old as the true silk coming from cocoons of cultivated silkworms fed on mulberry leaves. It is obtained from worms that have been raised on uncultivated mulberry leaves of the cherry tree, oak tree, castor oil plant, etc.. Tussah is spelled in a variety of ways—Tussa, Tassui, Talar. The silk comes mainly from China and India. Usually 18 cocoons are reeled together, the resultant thread averaging about (35.6 - 37.8 dtex) 32 ~ 34 denier in size. The amount of the silk production is limited because of the primitive conditions of its natural growth and weaving. Wild silk is strong, irregular, uneven. It is less lustrous than true silk. The fiber length is even longer than true silk: 600 yards (548m) up to about 1800 yards (1645m) or more.

化纤长丝纱 3.6 Man-made Filament Yarns

长丝类型 3.6.1 Filament Forms

As man-made fibers are sold on the market they may be in one of four different physical forms: monofilament, multifilament, staple and tow.

单丝 3.6.1.1 Monofilament

Yarns composed of only one fiber filament are called monofilament, in extruding these fibers, the hole in the spinnerette is the size needed for the final single filament yarn. Monofilaments tend to be somewhat stiff and lack the draping qualities of multifilament yarns. Filaments for use in either monofilament or multifilament form are very long. Actual length is limited only by the amount of yardage that can be handled on processing packages, which will be hundreds or thousands of yards in length.

复丝 3.6.1.2 Multifilament

When a number of individual filaments are combined to form a single yarn, it is identified as a multifilament yarn [Fig. 3 - 23(a)]. Multifilament yarns are made by extruding the desired number of filaments simultaneously through one spinnerette. The number of openings in the spinnerette is the same as the number of filaments to be combined into the final single yarn. (喷丝孔数)

The multifilaments are held together by twist or some other means, Twist is the normal procedure, and the amount of twist may vary from a very low to a high number of turns per inch.

短纤维 3.6.1.3 Staple Fiber

Man-made fibers may be cut into short lengths for processing on machinery used to cotton or wool. The length of the cut will be determined by the type of spinning equipment to be used. These cut fibers, varying in length from about 1 inch to 3 inches or more, are called staple fibers [Fig. 3 – 23(b)]. Before cutting the filaments are frequently crimped so that the staple fibers will have more cohesiveness. 卷曲/抱合力

Staple fibers are packed into bales and shipped to yarn manufacturers for conversion into yarns and then fabrics. Some staple fibers may be shipped directly to fabric manufacturers when nonwoven structures that do not require yarns are to be made.

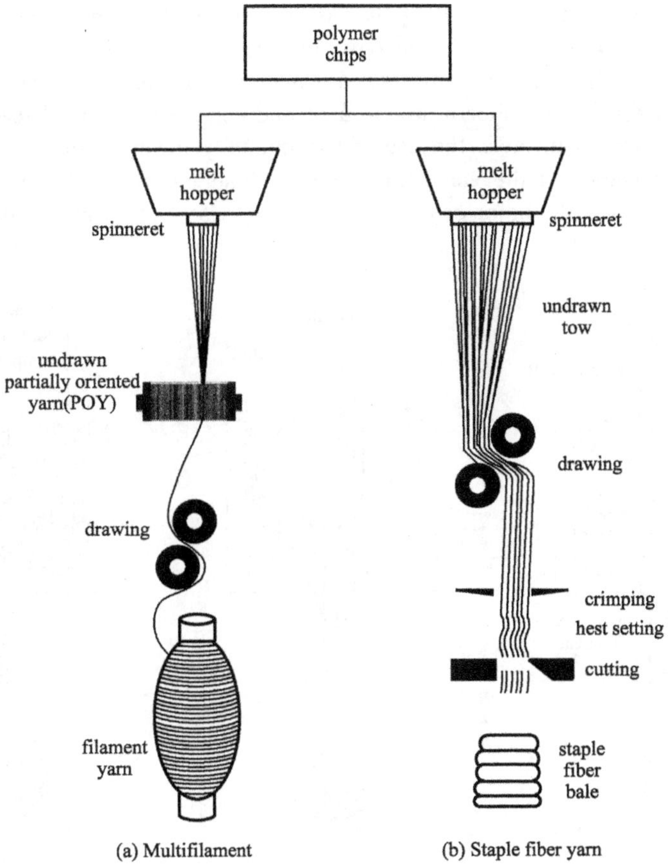

Fig. 3 – 23 Man-made filament and staple fiber

3.6.1.4 Tow

丝束

A considerable amount of man-made fiber is marketed today in a form called tow. This is composed of many hundreds, even thousands, of filaments which have been formed simultaneously by extrusion through special spinnerettes. These filaments are pulled together to form a large sliver of fibers. Manufacturers can process tow into yarns by a direct process or by an intermittent process. The direct process takes the sliver to special machinery that converts the tow into yarn as it comes from the fiber-forming units. In the intermittent process the tow is cut into short lengths and then processed as any staple fiber.

双组分和双成分长丝 3.6.1.5 Bicomponent and Biconstituent Filaments

Extruding two filaments of different composition so that they coalesce as they coagulate or harden produces bicomponent fibers, There are three basic methods for making bicomponent fibers.

并列型 (1) The side-by-side, S/S this type is made when two compatible filaments are spun side by side and they seal together as they emerge from the spinnerette. Special spinnerettes are required for all types of bicomponent fibers.

皮芯型 (2) The sheath-core, S/C process, in which one component is extruded in such a manner that it forms a sheath around a second component, called the core.

(3) The process involves the use of two mutually incompatible polymers that, when spun together in the same mix, form a filament in which one component forms fibrils within the other component, This type is called a matrix fiber, M/F. (Fig. 3 - 24). 基质型,海岛型纤维

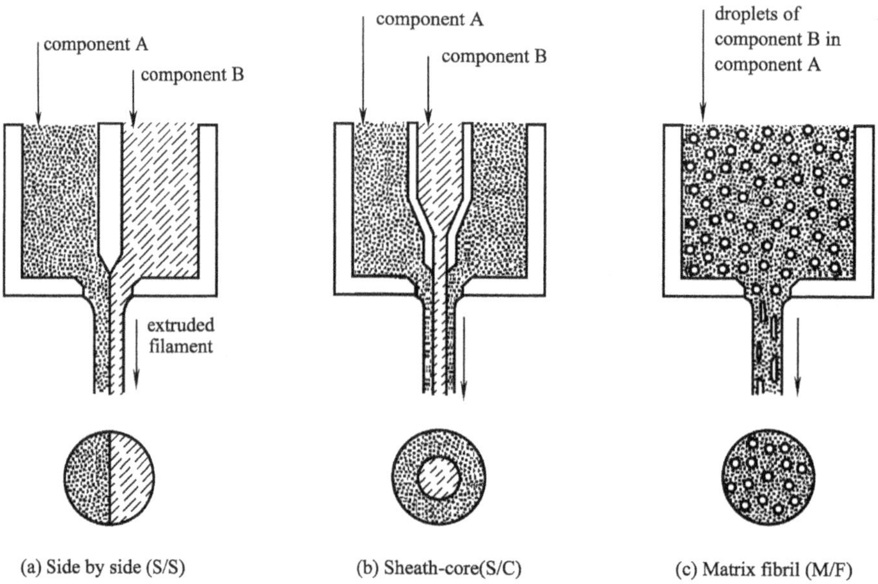

Fig. 3 - 24 Three types of bicomponent and/or biconstituent fibers. Extrusion techniques above; fiber cross sections below

Bicomponent fibers usually exhibit different shrinkage levels and different thermal behavior for each component. These differences enable yarn and fabric producers to introduce greater bulk, improved fit, retained sheerness, and attractive appearance into the end-use products.

Biconstituent filaments are formed when two totally different generic types of polymers are mixed together during or just prior to extrusion. These fibers have characteristics of both materials. Usually, biconstituent fibers are composed of distinctly different fiber types in order to maximize the characteristics of both.

3.6.2 Textured Yarns 变形纱

The term textured yarns for many years was just one of several expressions used in describing complex or novelty yarns. More recently, however, the term has been confined to a specific group of yarns and manufacturing processes.

Today's textured yarns are composed of filament or spun fibers. Most are processed directly from manufactured filament fibers. They may be regular or irregular in construction, so they resemble either simple or complex yarns (Fig. 3 - 25). However, textured yarns do possess individual properties of their own, and for this reason they require special attention.

Fig. 3 - 25 Textured (left) and nontextured yarns compared

ASTM defines textured yarns under the general heading of bulk yarns: 膨体纱

Bulk yarn: A yarn that has been prepared in such a way as to have greater covering power, or apparent volume, than that of conventional yarn of equal linear density and of the same basic material with normal twist. Varieties include bulky yarn, textured yarn, and stretch yarn. 弹力纱

Bulky yarn: A generic term for yarns formed from inherently bulky fibers such as man-made fibers that are hollow along part or all of their length, or for yarns 中空 formed from fibers that cannot be closely packed because of their cross-sectional shapes, fiber alignment, stiffness, resilience, or natural crimp.

Textured yarn: A generic term for filament or spun yarns that have been given notably greater apparent volume than conventional yarn of similar fiber (filament) count and linear density. The yarns have a relatively low elastic stretch. They are sufficiently stable to withstand normal yarn and fabric processing... and conditions of use by the ultimate consumer. The apparent increased volume is achieved through physical, chemical, or heat treatments or a combination of these.

Some authorities have used the term textured yarn to include all three types called bulk yarns by ASTM. This section deals primarily with yarns defined as either bulky or textured.

Most textured yarns are manufactured from thermoplastic fibers. The ability to be 热塑性纤维 heat-set, to be influenced in character by the application of controlled heat, is a 热定形

necessary property for the production of many, but not all, textured yarns. In recent years several manufacturers have developed finishing and treatment techniques that impact stretch in completed yarns. However, in this section discussion is confined to those textured yarns formed as a result of fiber manipulation.

填塞箱卷曲法 齿轮卷曲变形法/假编织法/假捻法/空气或流体变形法/刀口卷曲变形法

Three types of textured yarns are shown in Fig. 3 – 26. There are several processes used in the manufacture of textured yarns. These include the following: stuffer box crimping, gear crimping, knit-de-knit, false twisting, air-jet or fluid texturizing, and edge crimping.

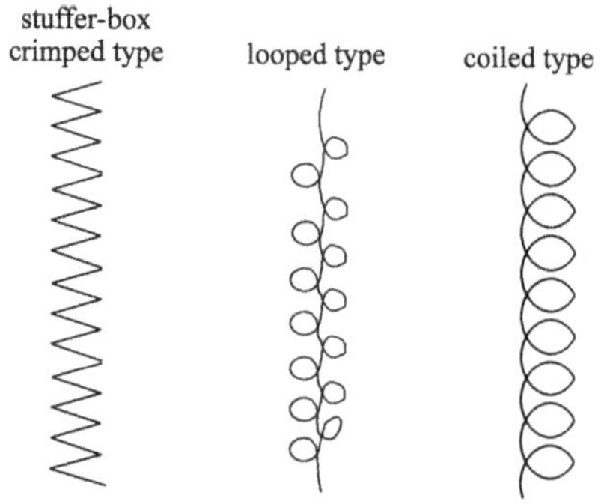

Fig. 3 – 26 Three types of textured yarns

Despite the fact that textured yarns can be manufactured from staple fibers, filament fiber yarns are preferred for texturing. These filament yarns are characterized by little or no pilling, yet they have the softness and bulkiness of staple fiber yarns.

3.6.3 Stretch Yarns

弹力纱

拉伸

The word stretch has acquired a specific meaning in modern terminology. A true stretch fabric has the ability to extend or stretch under tension plus the equally important capacity to return to its original size after release of strain. The degree of potential stretch, sometimes referred to as elongation, varies from as little as 5 percent to as much as 500 percent. Most stretch fabrics have an elongation or stretch factor between 10 to 25 percent for regular wear or 35 to 50 percent for fabrics designed for holding power. Stretch in these fabrics is expected to occur at low strain loads or low amount of pull.

弹性纤维

Core spinning is a procedure by which an elastic spandex filament is fed directly into the twisting zone of the spinning frame. The tension under which the elastomeric filament is held is carefully controlled. The resulting yarn has variable degrees of stretch, depending on manufacturing controls. Spandex filament forms the core of the yarn and is surrounded by staple fibers of one or more types. Core-spun yarns have two significant advantages, Only 3 to 10 percent spandex is required to produce a high-quality stretch yarn. Furthermore, the percentage of stretch can be controlled from 10 percent or less to as much as 200 percent. Most core-spun yarns are used in comfort stretch products and do not exceed 30 percent stretch at low loads. Core-spun yarns acquire the hand and appearance of the fibers used to form the outer sheath. Depending on the fiber used as covering, the fabric may have good moisture absorbency. The care of core-spun yarns is, basically, the same as that required of the covering fibers. 包覆纤维

Review Questions 思考题

1. Describe the carding process and what it does to the cotton fibers.
2. Describe the ring spinning process for making yarns.
3. Describe the various new yarn manufacturing systems.
4. What's the difference between open-end and conventional ring spinning?
5. Describe the rotor spinning process for making yarns.
6. Describe the friction spinning technique.
7. Compare carded yarns with combed yarns. How do they differ? How are they alike?
8. Define and explain the terms: filature, silk throwing, tram silk.
9. Explain the term of silk reeling.
10. What is meant by S-twist and Z-twist?
11. How does yarn twist affect the finished fabric?
12. Explain the difference among ply yarns, cabled yarns, and doubled yarns.
13. How do filament yarns differ from yarns of staple fibers in the manner in which they are made?
14. What are the main purposes of bulk yarns?
15. What would be the denier size of a yarn if 9,000 m of yarn weight 75 g?

Suggested Activities 实训题

1. Remove yarns from several different pieces of fabrics and determine if they are single, ply or cord.
2. If a laboratory is available, select yarn samples from various fabrics and measure to determine the yarn number, turns, or twists per cm of the yarns.
3. Visit stores and determine what types of products you can identify as having been made of textured yarns. Is this information easy to obtain?
4. Compare fabrics made of textured filament yarns with those that have not been textured. How do they differ? How are they alike?
5. Prepare rough diagrams to illustrate the processes involved in the manufacture of staple fiber and filament fiber yarns.

Chapter 4　Weaving and Woven Fabrics

机织与机织物

Weaving is one of the oldest arts known. Although no actual looms from early civilizations survive, fabrics of fine quality have been found in the tombs of ancient Egypt, and designs on very old pottery provide indisputable evidence of early skill in weaving. Painted pottery also gives us some idea of the looms used in the ancient world.

织机

Early looms were very crude compared with modern mechanical weaving contrivances. Nevertheless, all looms, old and new alike, have a few basic principles in common:

① There is some system to hold the long threads under tension.

线

② There is a way to spread yarns apart so that the crosswise thread can move through the opening or shed.

梭口

③ There is a device to pack the crosswise threads tightly together.

During the late eighteenth and early nineteenth centuries, developments in weaving involved the addition of automatic features to existing looms to increase the speed of operation and reduce the frequency and amount of damage due to faulty functioning.

Generally, the warp yarns undergo greater stress than the filling yarns during weaving; consequently, the yarn requirements for the two purposes are different. Warp yarns must be of a certain minimum strength, whereas the filling yarns can be quite weak; the warp yarns usually have a relatively high twist but twist in the filling yarns is usually kept as low as possible. As twist costs money and excessive twist produces harsh fabrics, it is generally kept as low as possible in both types of yarns.

经纱/纬纱

The present chapter covers the topics of yarn preparation for warp and filling, weaving and woven fabrics in detail.

4.1　Yarn Preparation for Weaving

织前纱线准备

Yarns as manufactured and packaged are not in the optimum condition to be used to form fabrics. After yarn formation, both spun and continuous filament yarns are not immediately usable in fabric forming systems. Package size, build and other factors make it necessary for the yarns to be further processed to prepare them to be handled efficiently during fabric formation.

织物成形系统/卷装/卷装形式

For weaving and warp knitting, many yarns are presented simultaneously in the form of a warp sheet. These yarns are taken from packages called beams. Shuttle looms need a special filling yarn package, or quill, which fits in the shuttle; while shuttleless looms and weft knitting machines use yarn from large packages called cheeses or cones.

经编
织轴/有梭织机
纡管/梭子
无梭织机/纬编
扁圆柱形筒子/锥形筒子

From the above it can be seen that the yarn, packaged as it comes from spinning, is virtually useless. It must be repackaged to meet the particular needs and demands of the fabric forming system in which it is to be used. This, in fact, is one of the functions of yarn preparation, to put the yarn on a suitable package for a particular fabric forming system.

Occasionally, it is necessary to alter some of the yarn characteristics to produce a yarn which can more easily and more efficiently be made into fabric or to produce a desired characteristic in the finished fabric. In this case, this operation would be part of yarn preparation.

A flow chart outlining the steps in the preparation of yarn for weaving is given in Fig. 4 – 1. It can be seen from this chart that, in terms of processes, a discussion of yarn preparation for weaving will, necessarily, include yarn preparation for knitting.

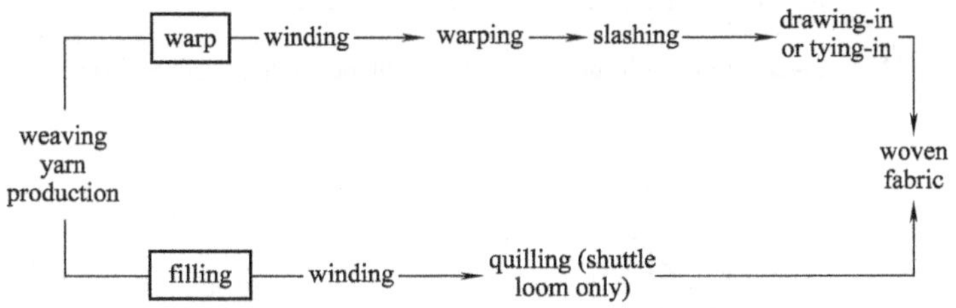

Fig. 4 – 1 Yarn preparation for weaving

4.1.1 Winding

络筒

The first step in yarn preparation for weaving is winding.

The reasons for winding yarn are:

① To produce a package which is suitable for further processing.

② To inspect and clear (remove thick and thin spots) the yarn. 粗节和细节

To perform the above tasks a winder, schematically illustrated in Fig. 4 – 2, is divided into three principal zones: 络筒机

① The unwinding zone. 退绕区

② The tension and clearing zone. 张力及清洁区

③ The winding zone. 卷绕区

To rewind the yarn on a new package, it must first be removed from the old package. This is accomplished in the unwinding zone. This zone merely consists of a creel, which holds the old package in an optimum position for unwinding. Fig. 4 – 3 illustrates the common yarn withdrawal methods, side withdrawal and over-end withdrawal.

筒子架
侧向退卷/轴向退卷

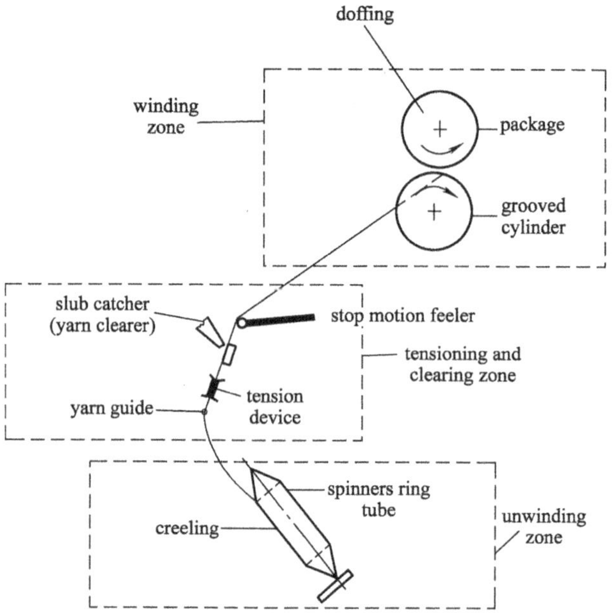

Fig. 4 – 2 Schematic of a typical winder position (friction drive)

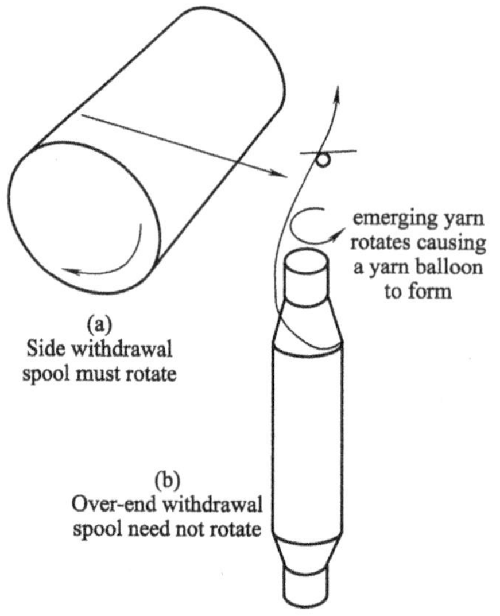

Fig. 4 – 3 Yarn withdrawal methods

In the side withdrawal method, the spool must rotate in order for the yarn to be removed. The advantage of this system is that the yarn does not rotate upon withdrawal and therefore the yarn twist remains constant. Its disadvantage is that the spool must rotate. At high winding speeds, due to inertia, the rotation of the spool may lead to tension variations in the yarn. Furthermore, provision must be made to stop the spool if, for any reason, the winder stops. If this is not done the rotational momentum of the spool will cause it to remain in motion allowing yarn to be unwound without being taken up. Also, upon start-up, higher tensions are developed because the winder must overcome spool inertia.

In the over-end withdrawal method, the package needs not be rotated as the yarn is pulled over the end of the package. This is the simplest and most common method of yarn withdrawal. There are, however, two factors which must be taken into account when this method of withdrawal is used.

The first of these factors is known as ballooning. As the yarn is unwound from the package at high speed, centrifugal force causes it to follow a curved path. As the yarn rotates, it gives the illusion of a balloon above the package. This ballooning leads to uneven tensions being produced in the yarn which may or may not alter some of the particular properties of the yarn. 气圈现象 离心力

The second factor for consideration is that for each time one complete wrap of yarn is removed from the supply package, the twist in that length changes by one turn. For most yarns this change is insignificant and may be ignored. However, some fabrics are constructed using flat yarns of metal, polymers or rubber. In these cases the yarn must remain flat and even one turn of twist is unacceptable. These yarns cannot be unwound using the over-end method and the side method must be used. 扁平纱线

The next zone is the tension and clearing zone. It is in this zone that the yarn receives the proper tension to provide an acceptable package density and build for further processing. This zone consists of a tension device, a device to detect thick spots, or slubs, in the yarn and a stop motion which causes the winding to stop in the case of a yarn break or the depletion of a supply package. The yarn is directed into this zone by a guide. 张力装置 粗节/自停装置 导纱器

Guides fall into two categories: closed, which require a yarn end to thread, and open, which do not. Open guides, however, give less positive guiding. It is important that guides be kept smooth to prevent damage to the yarn through abrasion, although friction damage can develop in too smooth a guide. Guides are usually made from hard stainless steels or from ceramics.

Following the guide the yarn enters a tension device. The purpose of the tension device is to allow the maintenance of proper tension in the yarn in order to achieve a uniform package density. The tension device also serves as a detector or excessively weak spots in the yarn which break under the added tension induced by the tension device. 探测器

Tension devices, as illustrated in Fig. 4 – 4, fall into three categories:

① Capstan (or multipicative) tensioner. 盘式张力器

② Additive tensioner. 倍积式张力器

③ Combined (or disc) tensioner. 联合式张力器

The most common type of tensioning device found on winding machines is the combined tensioner. This device consists of a capstan tensioner which accepts weight discs and thus also functions as an additive tensioner. The capstan is added primarily as a post-type yarn guide rather than a tensioning device and, in general, tension is regulated by adding or taking off the weight discs.

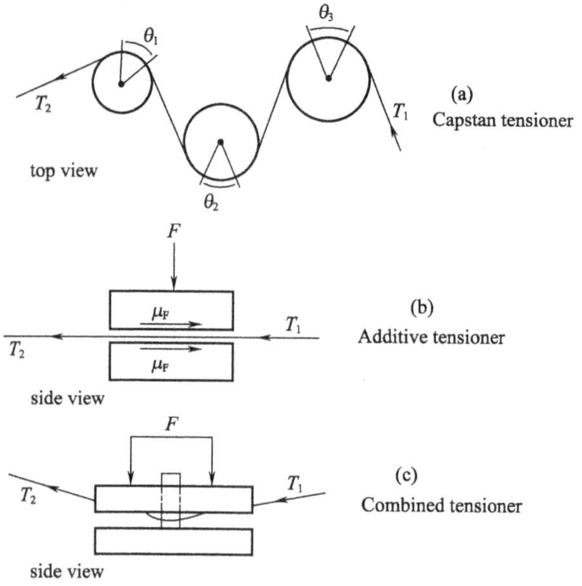

Fig. 4 – 4 Typical tension devices

Upon leaving the tension device, the yarn passes through a detector whose purpose is to detect thick spots. This detector may be as simple as a frame which contains an adjustable blade which can be set to allow only predetermined yarn diameters to pass through. This device is often called a snick blade. The detector, however, may contain sophisticated electronics which continuously monitor the yarn to detect thick (or thin) portions.

清纱板

After leaving the slub catcher, the yarn passes through a stop motion device. The purpose of the stop motion is to stop winding when the yarn breaks or runs out. This stop motion varies in configuration from machine to machine but in general consists of a counter-weighted or spring loaded sensing device which is held in an inactive position if the yarn is present. Breakage or running out causes the absence of this restraining yarn and allows the sensing device to activate.

重物平衡器/弹簧负荷传感器

The yarn is now ready to be put on a suitable package in the winding zone. This package may be one of many types, a cone, a tube, a cheese, a dye tube or a spool, depending upon the next operation the yarn must encounter. It is important that, during winding, no twist change take place. Thus physically wrapping the yarn around the package during winding should be avoided. The yarn is wound on the package by only rotating the package. This rotation may be accomplished in one of two ways:

锭子传动 ① Spindle drive, where the spindle upon which the package is placed is driven directly.

摩擦传动 ② Friction drive, where the spindle upon which the package is placed is free to rotate and the package is driven, through friction, by contact with a driven drum.

传动鼓轮

往复沟槽

A method of traverse found only on friction drive winders is the use of a traversing groove cut into the friction drum. In this method of traverse, illustrated in Fig. 4 – 5, the yarn rides in the groove in the friction drum and is carried back and forth along the length of the package. 摩擦鼓轮，槽筒

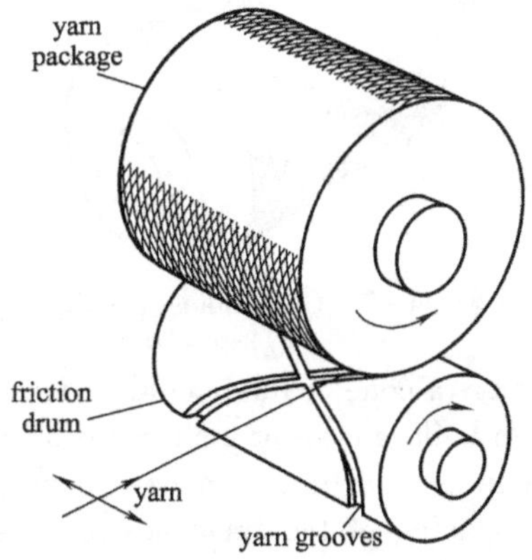

Fig. 4 – 5 Grooved cylinder traverse

All spindle drive winders and some friction drum winders use a reciprocating traverse, shown in Fig. 4 – 6, in which an externally driven guide carries the yarn back and forth across the package. The main advantage of this method of traverse is the ability to precisely lay the yarn onto the package. 往复横动

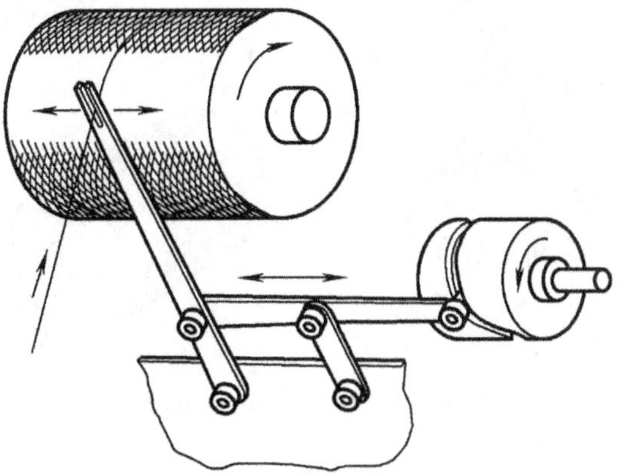

Fig. 4 – 6 Schematic diagram of a cam – driven reciprocating traverse mechanism

The type of package which may be built depends upon a combination of winding speed and traversing speed. If the traversing speed is relatively fast, successive layers of yarn will be laid at distinct angles to each other as shown in Fig. 4 – 7. This produces what is known as a cross-wound package. Because of the angle between successive yarn layers the shoulders of such a package are stable and do not 交叉卷绕

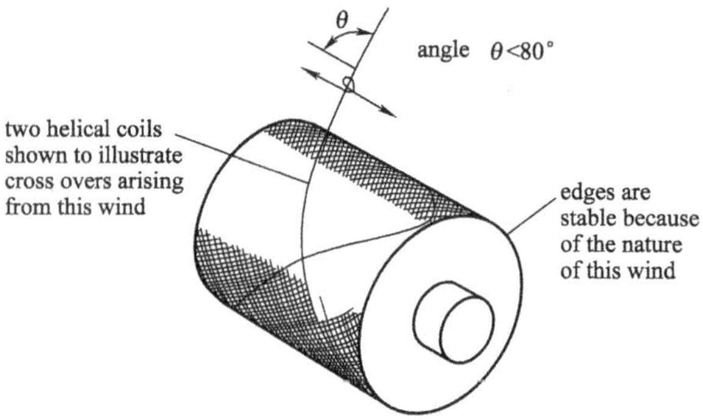

Fig. 4 – 7 Cross-wound package

need to be supported. Thus, a cone or tube could be used in the winding process. The traversing necessary to build a cone or a cheese is slightly different. Due to the difference in geometry between the tubular cheese package and the conical cone package, the angle and spacing of the traverse are constant in the case of the cheese but vary in the case of the cone.

If the traversing speed is relatively slow, successive layers will be very close to parallel to each other and a parallel-wound package will result. Parallel wound packages, illustrated in Fig. 4 – 8, are not stable and the shoulders of these packages need to be supported by flanges. Thus, for this type of wind, a spool is an appropriate package.

平行卷绕

边缘

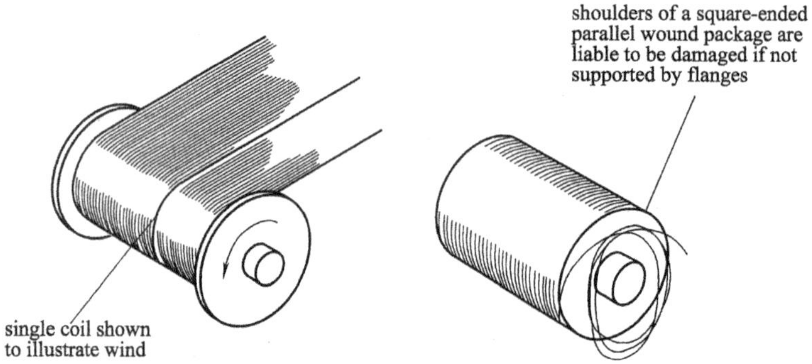

Fig. 4 – 8 Parallel-wound package

In applications where the package wind angle is important, such as yarn for weft knitting and filling for shuttleless weaving, it is important to ascertain and maintain a critical wind angle to prevent, or at least reduce, a condition wherein many coils of yarn unwind at a time from the package. This condition is known as sloughing-off. It is also important that the wind angle be such that the force required to remove the yarn remains constant.

临界卷绕角/纱圈

塌边

If the fabric design calls for yarn dyeing then the yarn is wound on a special tube which facilitates dye penetration into the package. After dyeing the yarn is normally rewound and sent to the next operation.

4.1.2 Quill Winding 卷纬

If the yarn is to be used as filling in shuttle looms it must be repackaged on a quill. The quill is designed to rest within the shuttle. There are spinning frames in existence which package filling yarn directly on quills rather than ring tubes. The productivity of this ring frame is limited and hence economics must be considered. To be more flexible, most mills use special winders designed specifically for the purpose of winding quills.

A schematic representation of a quill winder is given in Fig. 4 – 9. It may be seen in this figure that a quill winder differs very little from a package winder, the differences being no need for clearing and a different traverse mechanism. 卷纬机

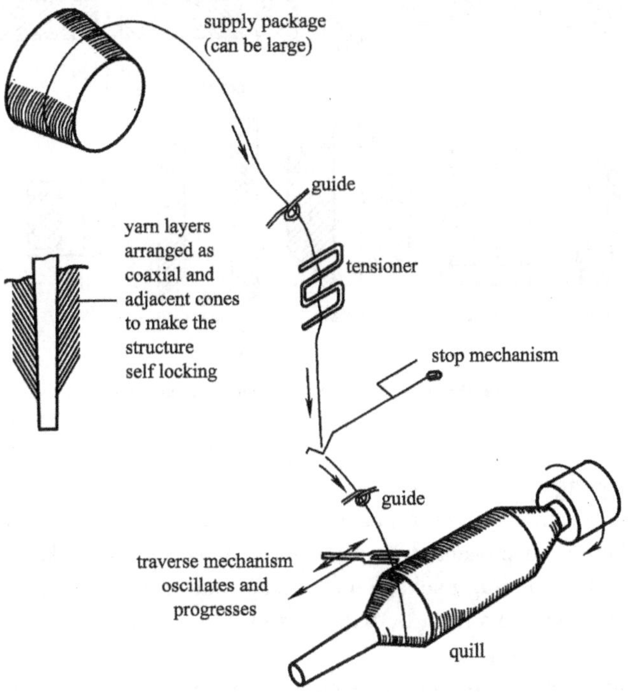

Fig. 4 – 9 Quill winder

Recall that, in package winding, the traversing mechanism makes a full cycle in carrying the yarn completely back and forth along the package. In quill winding, however, the traverse only covers part of the quill at a time. When one section is built up the traverse indexes to the next section. This is called building a quill by chase lengths or chasing a quill and the traverse is called a progressive reciprocating traverse. 往复卷绕长度

This method of quill winding is used for the following reasons:

① To help reduce the tendency to balloon as the yarn is unwound from the quill.

② To help maintain uniform tension in the filling yarns.

③ To reduce the possibility of sloughing-off.

If the quills are not to be used immediately after winding, they are usually taken to a room to be conditioned with hot, humid air. This conditioning is done to allow the filling yarn to relax, reducing the twist liveliness of the yarn and preventing the formation of kinks. 扭结

The type of package produced at winding varies with the needs of the next process. Some common winding packages are illustrated in Fig. 4 - 10.

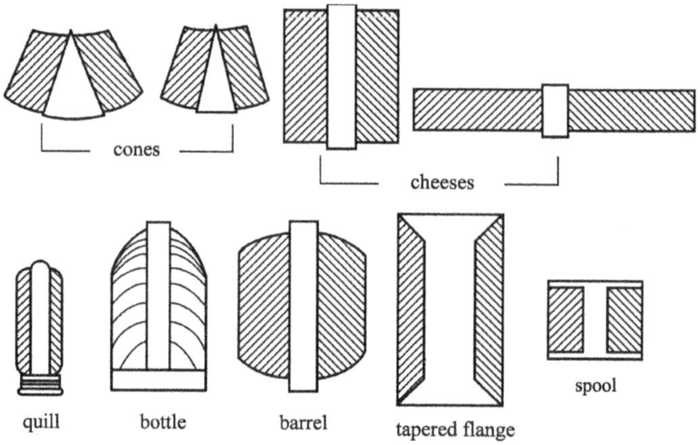

Fig. 4 - 10 Some typical winding packages

4.1.3 Warping 整经

If the fabric forming system is weaving or warp knitting, some or all of the yarns forming the fabric are presented in sheet form. It is necessary therefore to remove the yarns from the winding package and arrange the desired number on a package called a beam. The yarns must be parallel and under uniform tension. This, then, is the purpose of warping. 经轴

Before thinking about winding a specified number of yarns on a beam, first consider the problem of positioning the packages from which the yarn is taken in such a manner so as to facilitate the removal of yarn. Also keep in mind that the number of yarns per beam is in the hundreds or thousands and that there must, at least, be one supply package for each of these yarns.

It is logical, therefore, to build a frame of some sort to hold the packages. This frame is known as a creel and its function is to hold the supply packages in a manner so as to facilitate warping. To accomplish this purpose creels are equipped with package holders on which the supply packages are placed, tension devices to help maintain uniform tension throughout the creel, guides to direct the yarn and to help keep the ends apart, antistatic devices to eliminate static charges created by the rubbing of the yarn against the various surfaces and stop motions to detect broken ends and/or empty packages. 抗静电装置/静电荷 经停装置

4.1.3.1 Creels

筒子架

In theory, the size of the creel (and therefore the number of packages it may hold) is unlimited. In practice, and not considering purchase price, the creel size is limited by two factors.

The first of these is floor space. A creel must be housed in the building and therefore it necessarily uses some of the facilities of that building. Since the creel produces nothing tangible to offset the cost of housing and maintenance, it is important that it consume as little of these as possible. 有形资产/抵消

The second factor is the yarn itself. In theoretical discussions, yarn weight, especially for short lengths of yarn, is neglected. In considering a very large creel, it is obvious that some of the supply packages must be very much further away from the point where the beam is being formed than others. Also, the yarn must be supported to keep it from dragging on the floor and tangling. Each support acts as a capstan tension device. Thus, it is important to keep the packages in a distance range where the effect of yarn weight and the effect of supports as tensioners may be neglected. Hence the size and, therefore, the capacity of the creel is limited. 拖动/缠绕 张力器

In general, maximum creel capacity ranges from about 300 packages for very heavy yarns to 1,400 packages for fine yarns. As will be seen later, creel capacity is an important factor in warping.

Creels may be classified by the number of creel positions per end supplied. Using this classification, creels are either single or multiple package creels. 单式或复式筒子架

(1) Single Package Creels 单式筒子架

To achieve higher beaming efficiency, single package creels are often used in various combinations. If the winding head, or headstock, is fixed, often non-stationary single end creels are moved in and out of position as required. These creels are called truck creels. Truck creels as shown in Fig. 4 - 11 require that floor 车头/不固定的 车式筒子架

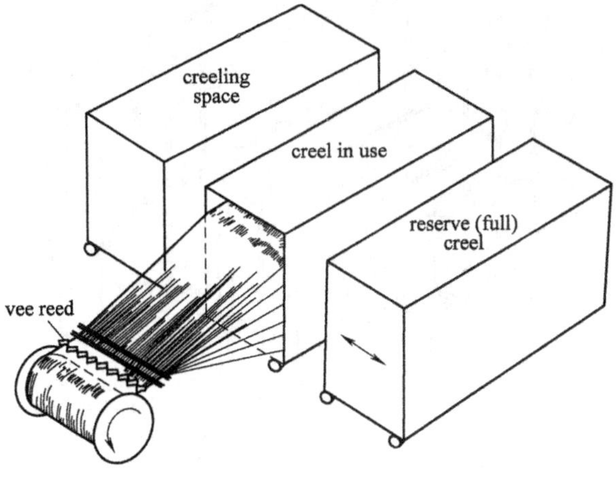

Fig. 4 - 11 Truck creels

space be reserved for the empty creel. A more space efficient set-up results if the headstock is capable of being moved. Creels used in this manner are known as duplicated creels, the advantage of which is the lack of need for an empty creel space.

双联筒子架

(2) Multiple Package Creels

复式筒子架

复式筒子架 In one type of multiple package creel, known as a magazine creel, illustrated schematically in Fig. 4 - 12, more than one package is provided for each end. The packages are tied head-to-tail so warping can continue when one package is exhausted.

头尾连接

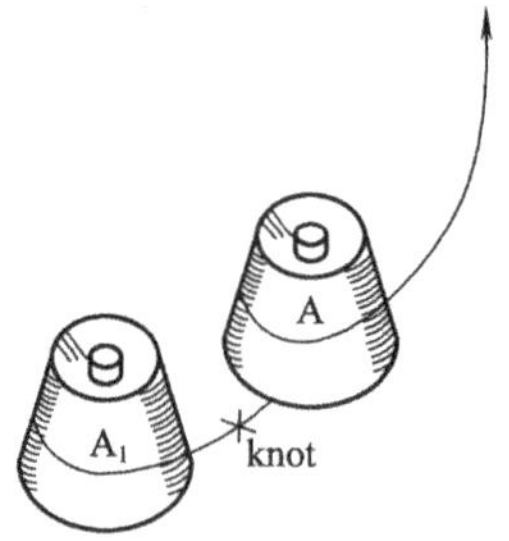

Fig. 4 - 12 Magazine creel

复式移行筒子架 Another multiple package creel, known as the traveling package creel, is illustrated by Fig. 4 - 13. Instead of moving creels or headstock when fresh packages are required, the packages themselves are moved into position. With a traveling package creel, the replacement of empty packages with full ones, or creeling, is done in the center while the packages in use are on the outside.

更换筒子

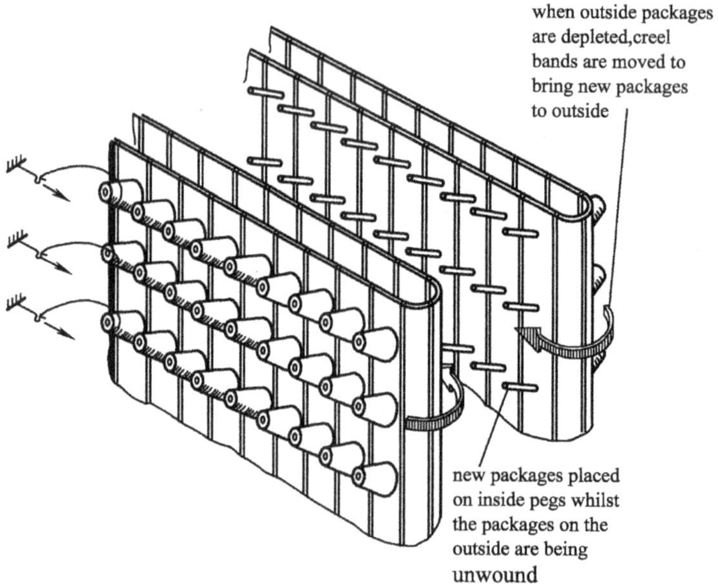

Fig. 4 - 13 Traveling package creel

4.1.3.2 Major Methods of Warping

整经的主要方法

Fig. 4 – 14 and Fig. 4 – 15 illustrate the major methods of warping. The warping is done depends upon the capacity of the creel, the number of ends required in the final beam and the necessity of maintaining a pattern in the warp, e. g. , for warp stripes in the fabric.

(1) Section Beam Warping

分条整经

If the creel capacity is sufficiently high and the total number of ends required is sufficiently low or, if creel capacity is not sufficient to supply all the required ends and no distinct yarn pattern is required, then beam warping is generally used. Beam warping (see Fig. 4 – 14) is simply the winding of yarns directly from the supply packages onto a beam. This beam is called a section beam since, except for the case in which all the required ends can be put on a single beam, it contains only a section of the warp required.

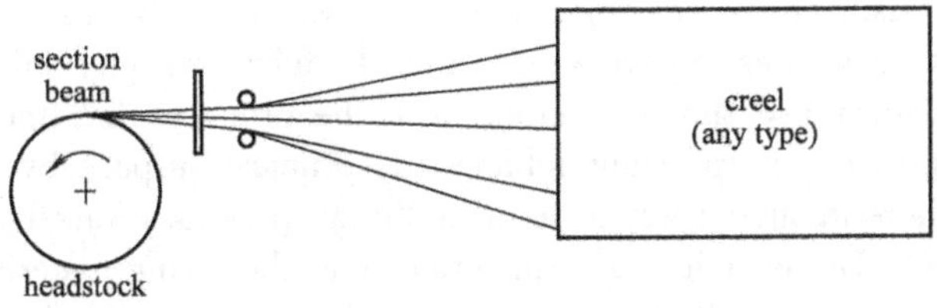

Fig. 4 – 14 Section beam warping

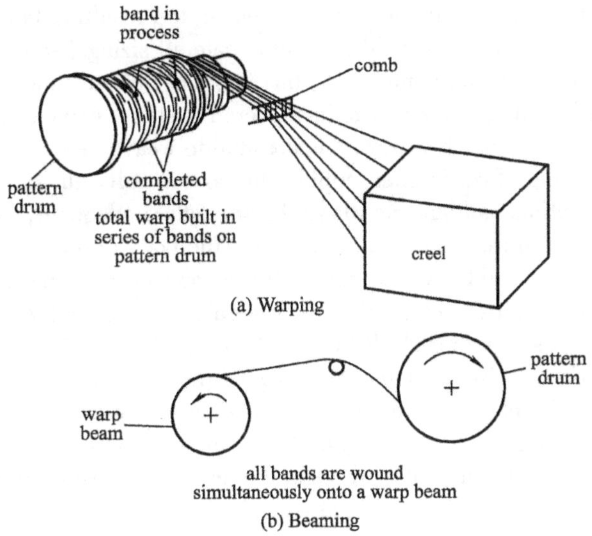

Fig. 4 – 15 Drum warping

(2) Drum Warping

大圆框整经

If, however, with insufficient creel capacity, it is necessary to build a warp beam containing the totality of ends required, or if the warp yarns have to be arranged in a definite order, then drum warping is used. In drum warping, the warp is not wound directly from the creel onto the beam but rather sections of the warp are wound onto a pattern drum, as illustrated in Fig. 4 – 15. In this manner the entire warp is built up in a series of sections on the pattern drum. When the total number of warp ends required in the fabric have been wound on the pattern drum, they are all removed simultaneously and wound upon a beam. This beam contains the exact number of ends required in the warp. Also, because when the ends are taken from the creel and wound on the pattern drum, exact placement in relation to each other may be made. The final beam maintains this placement, and hence any pattern in the warp.

条带 大圆框

浆经或浆纱 4.1.4 Slashing or Warp Sizing

概述 4.1.4.1 Introduction to Warp Sizing

The primary purpose of slashing or warp sizing is to produce a warp which will suffer the least damage in weaving. In some cases it is also used to modify the character of the yarn so as to have an effect on the fabric weight, stiffness or hand, but if this secondary use should interfere with the primary one, then the process has been misapplied. Warp sizing achieves its primary purpose by causing fibers mutually to adhere in such a way as to make the warp yarns stronger, smoother and better lubricated. However it is also important that the sizing materials should not interfere with the processes following weaving. Wherever possible the sizing should aid in these processes and not hinder them. Hence it is necessary to consider not only the way in which size is applied and its effects in weaving, but also the effects on subsequent processes (such as dyeing) and on the resulting fabric.

Although the ultimate aim should be to eliminate all sizing (and other preparatory processes) to achieve minimum cost, this is not at present practicable. Indeed, not only does adequate preparation reduce overall costs by making the weaving operation more efficient, but it is almost impossible to weave most warps found in industry without sizing. Size is often applied to yarns to give them added strength, but even with continuous filaments where the strength is adequate, lack of size may allow slack or broken filaments to protrude from the body of the yarn, especially in the cases of low twist and textured yarns. These protruding yarns can entangle or form fuzz balls and cause end breaks. In general, the higher the twist level, the less the amount of size required. If the twist is exceptionally high, it is possible to weave a warp without size, but the resulting fabric is harsh and unacceptable for most purposes. Acceptable fabrics may be obtained by weaving unsized ply yarns but the cost of plying has to be set against the cost of slashing. Usually slashing is used in preference to plying unless the fabric demands the use of plied yarns. Towels frequently fall into this latter category.

浆料 起球 未上浆的股线

Looking to the future, new yarn structures may modify sizing requirements. Twistless yarns (in which fibers are held together by an adhesive) obviously need little or no further sizing and it is quite possible that any attempt to slash them would cause yarn damage especially if the original adhesive were water soluble. Composite yarns (in which the fibers may be held in position by frictional or chemical forces and in which there are both staple and filament components and perhaps other polymeric material) may need special treatment which could differ quite radically from present experience. Open-end spun yarns (which have a more open structure in the outer sheath of the yarns) require a diluted size liquor. Furthermore, the weaving apparatus is undergoing change; shuttle-looms are giving way to shuttleless looms and the conditions imposed on the warp are altering. The water-jet loom, for example, is a special case in which the filling is inserted by a jet of water, and this obviously introduces sizing problems when a water-soluble size is used.

浆液

喷水织机

Water pollution has become a matter of increasing concern in our modern world, and there is mounting pressure to prevent industrial wastes (of which size is one) from contaminating rivers and other waterways. Consideration must be given to the question of recycling the material removed during desizing. One possibility is the use of reverse osmosis through membranes to separate the clean water and another is to use non-aqueous solvents in the size, but such solvents are either toxic or flammable and need great care to prevent hazards to the workers.

退浆

反渗透/膜

非水溶性溶剂

易燃的

The purpose of slashing (or warp sizing) therefore is to produce a warp which will withstand the rigors of weaving. This purpose is accomplished by:

① Enhancing the strength of the yarn by causing the fibers to adhere together.

② Making the outer surface of the yarn smooth.

③ Lubricating or waxing the yarn to reduce friction. 上蜡

In general, singles spun yarns must be slashed for all the above reasons. Continuous filament yarns, if they are not slashed at all, usually need adhesive to protect the filaments from breaking. Ply yarns are usually slashed for lubrication and/or smoothness. In general, size ingredients can be divided into four categories:

① Adhesives — Adhesives available include all types of starches, carboxmethylcellulose (CMC), polyvinyl alcohol (PVA) and others. 淀粉/羧甲基纤维素/聚乙烯醇

② Lubricants — Lubricants may be oils such as mineral and vegetable, waxes such as mineral, vegetable and animal or animal fats. 润滑剂

③ Additives — Additives may be included to provide features such as static elimination and mildew resistance. 助剂 防霉,抗霉

④ Solvent — The solvent generally used is water. 溶剂

The most important factors affecting the properties of sized yarn are:

① Lubricant added to fiber or filament prior to slashing (i. e., fiber finish or coning oil).

② Lubricant added to the adhesive as part of the size recipe. 浆液配方

③ Adhesive type.

④ Recipe (i. e., percentage adhesive, percentage lubricant, etc.).

⑤ Size add-on (the amount of size put on the yarn). 吸浆,吸液率

⑥ Technique of sizing.

⑦ Operational conditions such as yarn speed, temperature of drying, yarn tension etc..

⑧ Weaving room relative humidity.

The amount of size add-on depends on yarn structure and size recipe, thus there are significant differences between the requirements for flat filaments, textured and spun yarns. One major function of slashing is to control the surface of the yarn; in general, the more size add-on, the smoother the yarn and a smooth yarn will weave better than a hairy or non-smooth one. As the size add-on is increased the yarn becomes stiffer, more brittle, less extensible and more difficult to weave. At low size add-on, the yarn is more likely to break because of lack of strength and increased hairiness. The net result is that there is an optimum add-on level at which there is a minimum end-breakage rate and the best add-on level varies according to the smoothness of the yarn. For most spun yarns, depending upon fiber type and size recipe, the minimum value of warp breaks falls in the 5 % –15% size by weight range.

浆纱机
4.1.4.2 Sizing Machines

直接浆纱/间接浆纱
Fig. 4 – 16 illustrates the warp sizing machines of direct and indirect slashing. The direct and indirect methods are used frequently for spun yarns, although it is possible to slash textured yarns also if proper care is taken. In both the direct and indirect systems, beams are used to supply a sheet of yarn to the size box, and in the indirect case a beam creel is used and there are several possible beam arrangements. Negative let-off is normally used because of its simplicity, but occasionally the beams are directly driven and controlled to give constant warp tension.

浆槽 经轴架

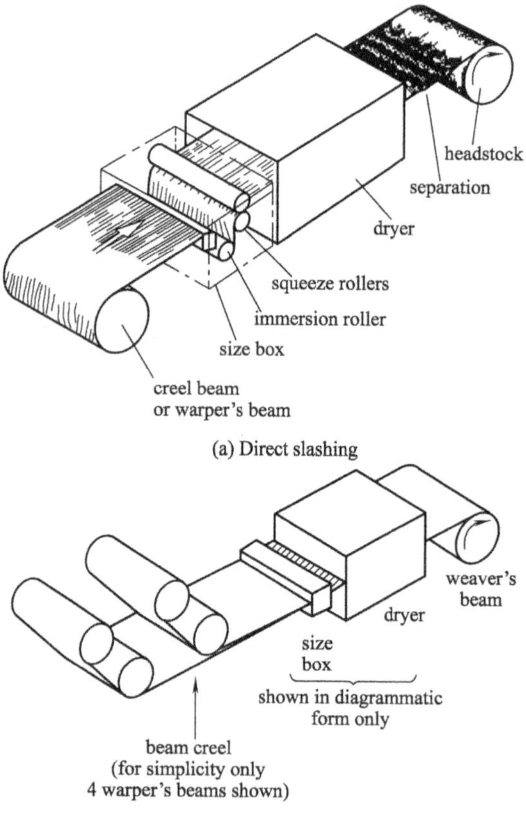

(a) Direct slashing

(b) Indirect slashing(use of section beams)

Fig. 4 – 16 Warp sizing machines

In general, warp sizing machines or slashers can be divided into five different sections: 浆纱机

① Beam creel.
② Size box.
③ Drying section. 烘燥区
④ Yarn separation section. 纱线分绞区
⑤ Headstock. 车头

The beam creel is merely a device or frame on which warp beams are placed in a manner convenient for unwinding. The creel can hold as few as one beam and, usually, as many as fourteen. Reflecting back to the packages produced in warping, recall that, in the case of beam warping, there were usually a number of section beams prepared, each containing a portion of the total number of warp ends required. The creel of the slasher holds all these beams and the ends on them are combined during the sizing operation. At the slasher, many warp beams are combined to form a single weaver's beam [Fig. 4 – 16(b)]. Remembering also that drum warping yielded a warp beam containing the totality of ends required for weaving, it is seen that, for this beam, no combination is needed at the slasher [Fig. 4 – 16(a)]. Often, in the case of multiple section beams and prior to the yarn's entering the size box, the yarn passes over a series of rods called lease rods. The lease rods help the yarns coming from many different warp beams to flow together smoothly. 织轴 多个经轴 绞棒

The yarn next enters the size box, shown in Fig. 4 – 17. The size box contains the size solution, known as size liquor. The yarn is fed into the size box by means of a guide roll. It then passes under a dip or immersion roll. This roll is capable of being moved up or down allowing the yarn to be held in the size liquor for a desired 导辊/浸没辊

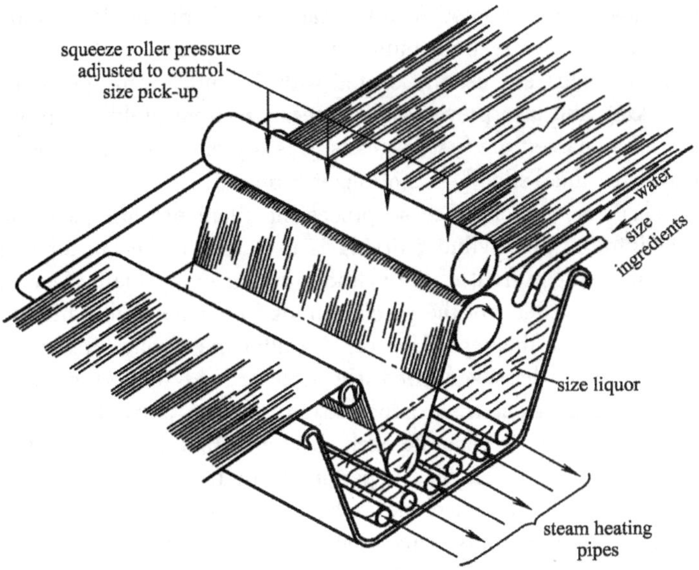

Fig. 4 – 17 Section of a size box

压浆辊 period of time. The warp sheet then passes through two rolls known as squeeze rolls. The purpose of the squeeze rolls is two-fold:

① To squeeze out excess size.

② To physically drive the size into the yarn for proper penetration.

The percentage of size by weight is controlled by the yarn's exposure to the size solution, governed by the speed of the machine and the immersion roll depth, the yarn structure and the pressure applied by the squeeze rolls.

The size box also contains pipes which supply solvent, size ingredients and steam to heat the size liquor. In this way, the temperature and concentration, and hence the viscosity of the size is kept as constant as possible to assure both correct and uniform size pick up by the warp yarns.

After the yarns have been exposed to the size liquor and have picked up the required amount of size, the size solvent must be driven off, i. e. , the yarns must be dried. This drying may be done by exposing the yarns to hot air, by passing them over heated cylinders (or cans), by exposing them to infrared or by a combination of these methods. 烘筒/红外线

生产率 The drying section determines the maximum throughput rate. It is required to dry the wet sized yarn rapidly, thoroughly and uniformly. A simple 2-cylinder machine is too slow as it is difficult to get a sufficiently high heat transfer rate. By introducing more cylinders, more drying surface is made available and the contact time for a given yarn speed is increased.

The most common drying method is the use of heated cylinders or cans. The yarn is dried by coming in contact with these hot cans. Residence time on the cans is important in order to dry the warp sheet efficiently. However, high speed may be achieved by increasing the number of drying cans, thus effectively increasing residence time. If the cylinder drying method is used precautions must be taken so the size will not cause the yarn to stick to the cans as it is drying. Precautions must be made to prevent excess size from building up on the cans. For these reasons. the first three or four cans are usually coated with Teflon® . This prevents the sized yarn from sticking and also prevents the excess size from building up on the cans. After the yarn has passed the first few cans, it is dry enough so as to no longer stick to noncoated cans and Teflon® is no longer required. 涂层/聚四氟乙烯

On some slashers, especially those processing filament yarns, a combination of the hot air or infrared and cylinder drying is used. Often the hot air or infrared is used in a pre-dry unit in front of the cylinder drying unit. This is done to minimize any possible negative effects of the yarn's coming in contact with a high temperature drying unit and to ease the separation of yarn after the drying section. 热风

After the sized yarn has been dried it is still not ready to be put on a beam; recall that one size ingredient is an adhesive. In order to prevent adhesion between the yarns, it is necessary to separate each sized end from the others before the warp can be used for weaving. This task is performed by the burst rods. These are positioned on the slasher and the machine is threaded in such a way that alternate ends are sent in alternate directions. This process may be repeated as many times as thought necessary to achieve total separation. Often, in the case of continuous filament yarns, there is a splitting section prior to the drying section. This is done in an 上浆的经纱 分绞棒 裂开区

effort to reduce any effects to the yarns from the initial shock of bursting a completely dried warp sheet.

The yarn is now ready to be put on the loom beam. It is threaded through an expansion or "zig-zag" comb which is adjusted to allow the warp sheet to come to the width required to fill the loom beam. Leasing facilitates the subsequent drawing-in or tying-in operations. 锯齿形 穿经 结经

4.1.5 Drawing-in and Tying-in 穿经与结经

4.1.5.1 Drawing-in 穿经

Drawing-in is the process of providing each end with a drop wire, a heddle in the proper harness and a reed dent in the reed as shown in Fig. 4 – 18. Drawing-in does not generally occur on the loom, but rather in another area in the mill. Drawing-in can be performed manually or by means of automatic machines. 经停片/综丝 综框/筘齿/钢筘

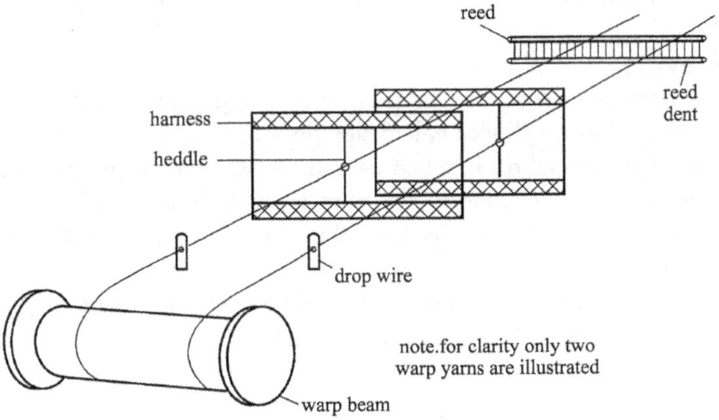

Fig. 4 – 18 Typical warp yarn path through a loom

4.1.5.2 Tying-in 结经

Tying-in is used when a fabric is being mass produced. The tail end of the warp from the exhausted warp beam is tied to the beginning of the new warp. Two types of machine are used:

① Stationary machines. The tying-in takes place in a separate room away from the loom. 固定式结经机

② Portable machines. These are used at the loom. 手提式结经机

Stationary machines have the disadvantage that they necessitate moving the exhausted beam and all its parts from the loom and taking it to and from the tying-in department. However, they have the advantage of permitting maintenance of the loom to be carried out.

The time taken to tie-in a complete warp sheets depends mostly upon the total number of ends in that warp, but it is also affected by secondary factors which tend to retard productivity. For example, a color stripe must be tied in proper register and the operator will have to stop tying if there has been a broken end in order to 接头,结经

adjust the machine to give proper register. The count and type of yarn (together with the reed and heddle details) determine the type of knot to be used and this affects the rate of knotting. Also the nature of the yarn can affect the breakage rate during knotting, and thus influence the total time needed for tying-in.

结经机 The capacity of warp-tying machines has remained unchanged for years. A capacity of about 600 knots appears to be the maximum. The machine can deal with

未分绞经纱/已分绞经纱 flat warp or leased warp and with a warp width of about 5m. The sequence of operations is normally as follows:

① The machine selects the warp ends from the new beam.

② It selects the corresponding end from the old beam.

③ It ties the two ends together and moves to the next.

Following the tying-in process, all knots are pulled through to the cloth roller, the drop wires, heddles and reed; the loom is now ready for operation. A similar process can be used where similar (but not identical) fabrics are to be produced. Obviously, identical drawing-in draft and reed plans arc required.

机织 4.2 Weaving

织轴 On the conventional loom, the warp beam is mounted at the back and the warp

卷布辊 yarns are conveyed to a cylinder called the cloth roll, which is at the front of the loom and on which the fabric is rolled as it is constructed. Supported on the loom frame between these two cylinders (the warp beam and the cloth beam), the warp

交织 yarns are ready to be interlaced by the filling yarns that run in the width of the cloth, thus producing the woven fabric.

In order to give some reality to these generalities, the schematic diagram of Fig. 4-19 shows the passage of the warp through a loom.

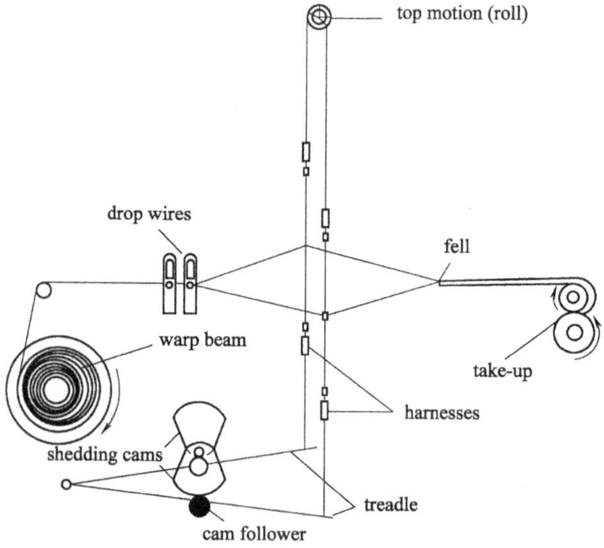

Fig. 4-19 Across section through the loom

The yarn from the warp beam passes round the back rest and comes forward through the drop wires of the warp stop-motion to the healds, which are responsible for separating the warp sheet for the purpose of shed formation. It then passes through the reed, which holds the threads at uniform spacing and is responsible for beating-up the weft thread that has been left in the triangular warp shed formed by the two warp sheets and the reed. Temples hold the cloth firm at the fell to assist in the formation of a uniform fabric, which then passes over the front rest, round the take-up roller, and onto the cloth roller.

后梁
自停装置
打纬
边撑/织口
胸梁
卷取辊

The mechanisms of a power-driven loom receive their motion from shafts that traverse from side to side in the loom and are driven from a motor. Their relative speeds are of importance since they govern the mechanisms that they drive.

机械传动织机,电力织机

4.2.1 Loom Basic Mechanisms

织机的基本运动

Today high-speed power looms are used to produce quality woven fabrics. In order to continuously produce a woven fabric on a power loom, four operations, known as basic or primary motions, must be done. The four primary motions are shedding, filling insertion, beat-up and warp and fabric control or letting off and taking up.

基本运动
开口/引纬/打纬/送经/卷取

4.2.1.1 Shedding

开口

Shedding is defined as the movement of the warp yarns to provide an insertion path for the weft or filling yarn(see Fig. 4 – 20).

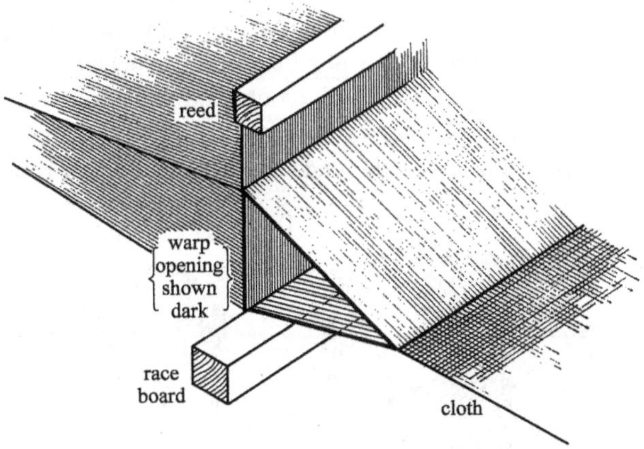

Fig. 4 – 20 Race board and warp shed

During drawing-in, each warp yarn is threaded through a heddle eye and, in turn, each heddle is assigned to a harness or, as will be seen, may individually act as a harness. It has previously been shown that the woven fabric design is dependent upon this assignment and to form a woven fabric there must be at least two distinct assignments. The formation of the shed is accomplished by the movement, lifting or lowering, of the harnesses.

综眼

凸轮开口系统 (1) Cam System

踏板 Early non-mechanical looms had these harnesses connected to treadles which the weaver could either step on or release to provide harness movement. Early modern 凸轮 looms replaced the weaver's foot with the mechanical device known as a cam. As shown in Fig. 4 – 19 the rotation of these cams is responsible for moving the harnesses, thus opening and closing the shed.

The cam system of shedding is extremely simple in both concept and method. 偏心 Through their eccentricity, the cams provide the power necessary to move the harnesses. The design in the fabric, caused by the movement of the harnesses through one design repeat, is controlled by the cam shape, or profile. Because of this dual role, the complexity of designs which may be produced on a cam loom is generally limited to designs repeating on six or less picks. The vast majority of woven fabrics fall into this range so the limitation is not as severe as might be expected. However, there are fabrics whose design is beyond the capability of a cam loom and so other methods of shedding must be used.

多臂开口系统 (2) Dobby System

The next major shedding system, the dobby system, increases its pattern capability by separation of the functions of harness movement and pattern control. The 多臂龙头 Fig. 4 – 21 shows a shuttle loom with a dobby attachment, Fig. 4 – 22 illustrates 吊钩 the method of movement in the dobby system. A harness is lifted when a hook associated with that harness linkage is positioned in such a way as to be caught by a 拉刀 knife. The knives are driven by the loom and, except for providing harness movement, play no part in the positioning of the hooks to generate the harness movement necessary for a particular fabric construction. The positioning of the hooks is 纹板链 accomplished independently by a mechanical action controlled by a pattern chain. 纹钉板 The pattern chain usually consists of a peg board or a punched cylinder of paper or plastic. The length of the design repeat is controlled by the length of the pattern chain which, at least in theory, may be infinite. The limiting factor in the design 经纱 capability of the dobby loom is that the ends are in harnesses and therefore the de-

Fig. 4 – 21 A typical modern shuttle-type loom with a dobby attachment

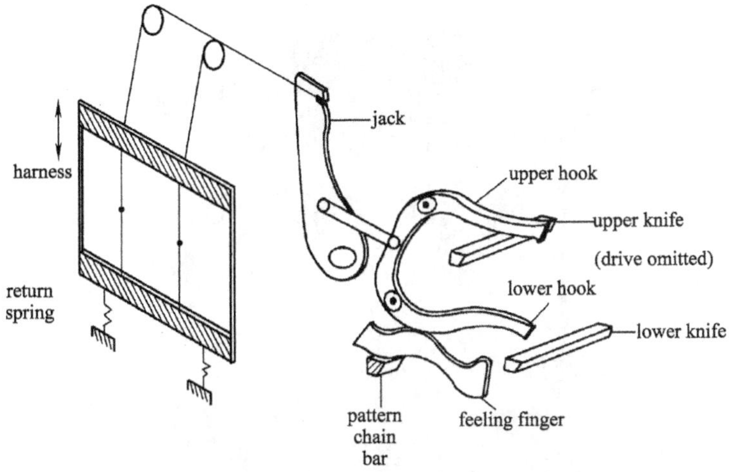

Fig. 4 – 22 Dobby shedding

sign may contain only as many distinct end movements, however intricate, as there are harnesses. In some very complex fabrics virtually every end must be capable of distinct movement and so an even more complex shedding system is required.

(3) Jacquard System 提花开口系统

In the third major system, the Jacquard system, a knife-hook system using the same principle as the dobby system, causes the lifting or lowering of each end. In this case however, the hooks are not associated with harnesses, but rather with individual ends or, at most, a small group of ends. Pattern control, accomplished by the positioning of the individual hooks, is done through the use of a continuous deck of punched cards or, in some new Jacquards, by endless tapes. In this way, maximum design capability is achieved in that virtually each end is capable of individual movement and, theoretically, the repeat size is infinite. Fig. 4 – 23 shows a Jacquard loom. Jacquard shedding is shown in Fig. 4 – 24. It is easily seen that the dobby and Jacquard systems are more alike than different and both, in many respects, are primitive digital computers.

提刀系统
提花机

With minor exceptions, all loom types are capable of having one of the above methods of shedding. Therefore the first major classification of looms is determined by the shedding type.

4.2.1.2 Filling Insertion (or Picking) 引纬,投纬

Filling insertion is the passing of the filling yarn through the open shed as shown in Fig. 4 – 25. While the traditional method of filling insertion is by means of a shuttle (see Fig. 4 – 26), new developments in weaving have centered around replacing the shuttle with other filling carriers. Looms using a filling carrier other than a shuttle are grouped under the general heading of shuttleless looms. The common filling insertion methods in use today include the conventional shuttle, rapiers, gripper projectiles, air jets and water jets, the last four being shuttleless systems.

载纬器
剑杆/片梭
喷气/喷水

Fig. 4 – 23 A Jacquard loom

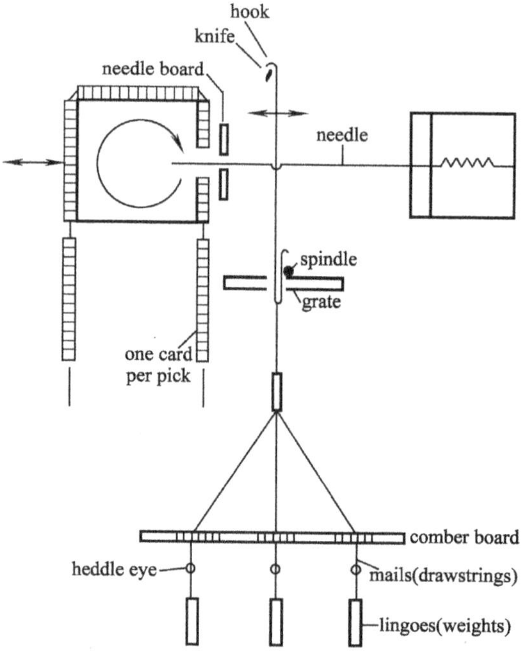

Fig. 4 – 24 Schematic Jacquard shedding

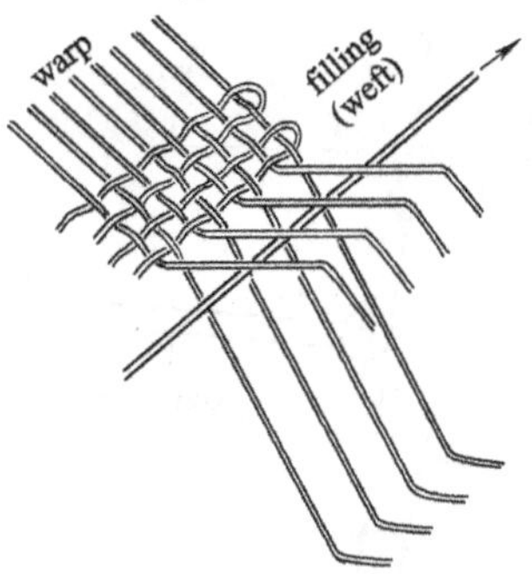

Fig. 4 – 25 Filling (weft) the loom

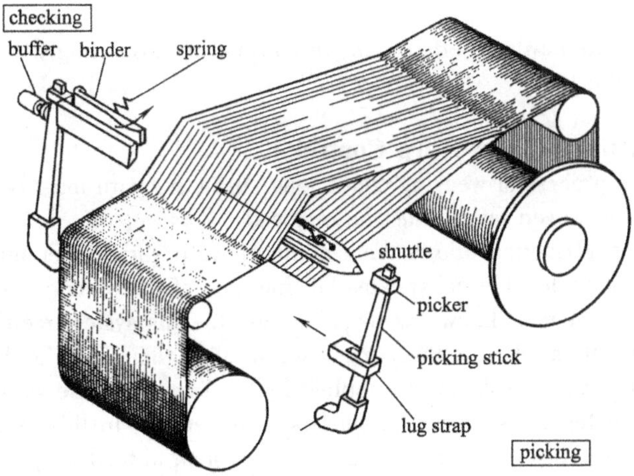

Fig. 4 – 26 Picking for a shuttle loom

The means of filling insertion provides the second major loom classification system. Therefore to completely categorize a loom, both shedding type and filling insertion method should be given.

4.2.1.3 Beat-up (Beating-up) 打纬

After the filling is inserted, by whatever means of insertion, it must be incorporated into the body of the fabric. This is the purpose of beat-up as shown by Fig. 4 – 27.

Beat-up is accomplished by the use of wire grate called a reed. The warp yarns are threaded through the spaces of the reed called dents. At the beginning of the weaving cycle, the reed is moved backward to allow for filling insertion. After the 织造循环
filling is inserted, the reed moves forward and the wires engage the filling yarn driving it into the fell of the cloth, the position in the warp shed where each pick is 每一纬
beaten. The usual mechanism which gives the reed its motion is known as the lay. 筘座

The main functions of the reed are to hold the warp threads at uniform spacing

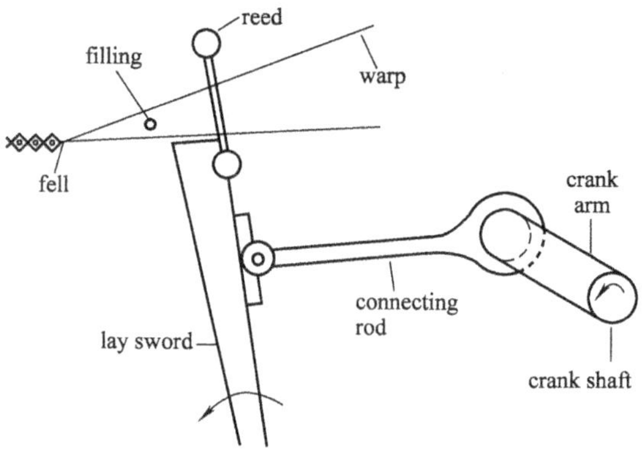

Fig. 4 – 27 Schematic beat-up mechanism for a shuttle

and to beat-up the newly inserted picks in addition to supporting the shuttle during its traverse of the loom.

送经(即经纱控制)

4.2.1.4 Letting off (or Warp Control)

To make the process of weaving continuous, the warp yarn must be continuously supplied and the woven fabric must be removed.

The warp let-off motion allows the feeding of warp yarn under uniform tension and at the proper rate. Let-off systems fall into one of three major categories. The simplest let-off system is known as negative let-off (消极式送经). This system merely consists of a brake continually applied to the warp beam to retard (阻止) the motion of the beam. The force required to turn the beam is supplied by the tension in the warp yarns.

The second category is known as semi-positive let-off (半积极式送经). In this system the loom physically drives the warp beam through a gear system to feed warp yarns at a predetermined rate. The let-off may be activated intermittently (间歇地) or continuously. Together, negative and semi-positive let-off systems make up the majority of these found on looms.

The third system is known as positive let-off (积极式送经). This system is primarily used to produce fabrics in which some of the warp yarns must be fed at different rates, for example, terry cloth.

卷取(即织物控制)

4.2.1.5 Taking up (or Fabric Control)

There are two basic take-up systems, the direct and indirect.

In the direct system the fabric is taken up and stored by wrapping it around a single beam called the cloth roll (布轴). To remove the fabric, i. e. , to doff the cloth roll (落布), the loom must be stopped.

The other method of fabric take-up is to have a set of rolls do the actual fabric take-up and have the cloth roll merely serve to wind up the fabric. This system is known as indirect fabric take-up (织物间接卷取).

It is also necessary to hold the fabric at the fell of the cloth to insure that the reed beats the filling across the full width (整个幅宽) of the fabric as nearly parallel as possible. This lateral fabric control is the function of the temples. The temples are simply mechanical fingers which maintain the fabric at full width at the fell of the cloth. 幅撑

4.2.2 Shuttleless Looms 无梭织机

Loom designers have constantly sought to replace the shuttle as a means of filling insertion. The prime reason is rate of production. Physical laws dictate that, to move through the shed, the shuttle must be accelerated and decelerated. This naturally takes time and energy. If less time could be spent accelerating the shuttle, and the transit time of the shuttle through the shed could be reduced, higher loom speeds could be obtained. Secondly, with the removal of the filling supply package from the filling carrier, the filling carrier could be made smaller so the yarn movement in shedding could be reduced.

Some features of shuttleless looms are:

① High rate of filling insertion.

② Use of large, exterior filling packages.

③ Limitation, to some degree, on yarn type.

④ Require additional hardware for selvage modification and filling accumulation systems. 储纬系统

⑤ Reduction of noise and vibration.

Not all of these features are positive, nor are they applicable to every type of shuttleless loom. Hence, on most common shuttleless looms, yarn type, for example, is not limited. We will examine the basic types of shuttleless looms (see Fig. 4-28) and briefly discuss each. Shuttleless loom technology has proceeded in two

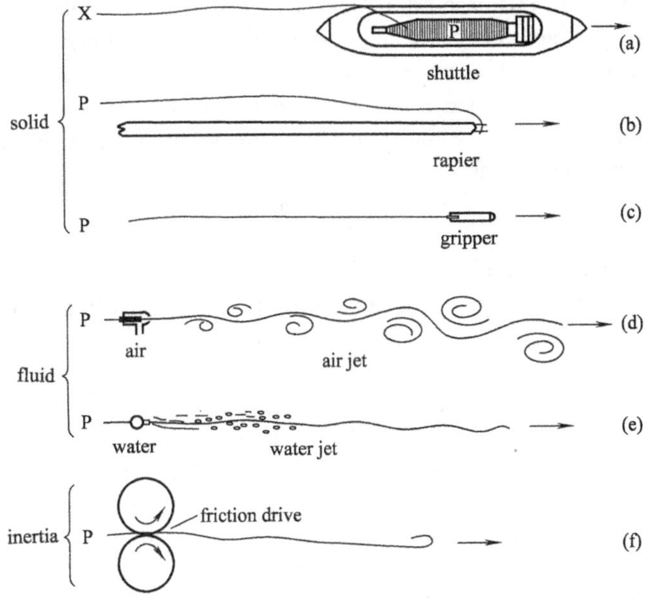

Fig. 4-28 Filling insertion systems

general directions in its attempt to replace the shuttle. In the first, as found in ra-
流体喷射织机 pier and gripper looms, the shuttle is replaced by another solid carrier. Fluid jet
looms, in which a high pressure jet of air or water is used to carry the filling
through the shed, represent the second direction.

剑杆织机 4.2.2.1 Rapier Looms

A rapier is merely a device made of metal or a composite material with an at-
tachment on the end to carry the filling yarn through the shed. The simplest ra-
单刚性剑杆 pier loom is known as the single rigid rapier. This type of loom consists of one
rapier which carries the yarn through the shed. In most instances this loom is
抽出 not as fast as a shuttle loom because the rapier must be withdrawn from the shed
prior to beat-up and, during this withdrawal, no filling is carried. Recall that
the shuttle loom lays a filling yarn with every pass through the shed. There are
also space considerations, since the single rapier must be at least as long as the
loom width.

双刚性剑杆 To overcome some of the speed limitations, the double rigid rapier loom was de-
送纬剑 veloped. This loom consists of two rapiers—one called a "giver" and the other a
接纬剑 "taker". The giver picks up the filling yarn and carries it to the center of the
shed. Meanwhile, the taker moves to the center of the shed, picks up the yarn
from the giver, and carries it the remaining distance. Now each rapier covers only
half the distance and therefore almost twice the number of picks per unit time can
be inserted as compared to the single rapier. Unfortunately, because of the rigid
rapier, there still may exist the problem of wasted space.

双挠性剑杆 To overcome the potential space problem, the double flexible rapier was devel-
oped. The rapiers in this case are no longer rigid bars but flexible tapes. Since the
tapes are flexible they can be bent so as to fit under the loom, thus conserving
space. The double flexible rapier consists of a giver and a taker similar to the
伸缩剑杆 double rigid rapier. Other looms use telescopic rapiers which extend during filling
insertion and retract during withdrawal, thus attempting to combine the features of
both flexible and rigid rapiers.

The different types of rapiers are illustrated in Fig 4 – 29.

片梭织机 4.2.2.2 Projectile (Gripper) Looms

The projectile loom uses a small projectile to carry the filling yarn through the
shed. In reality, the shuttle is actually a projectile but the distinctive feature of
shuttleless looms that the filling package be exterior to the filling carrier, dis-
qualifies the shuttle loom from consideration. The projectile loom uses a projec-
tile with grippers to carry the yarn through the shed. This projectile may be
picked from both ends or from one end. If picking is only from one end, there
must be a method to carry the projectile back to the other side of the loom,
瑞士苏尔寿织机 e. g. in the Sulzer loom, many projectiles are used and, combined with a con-
veyor belt return, give a high rate of filling insertion. These types are illustra-
ted in Fig. 4 – 30.

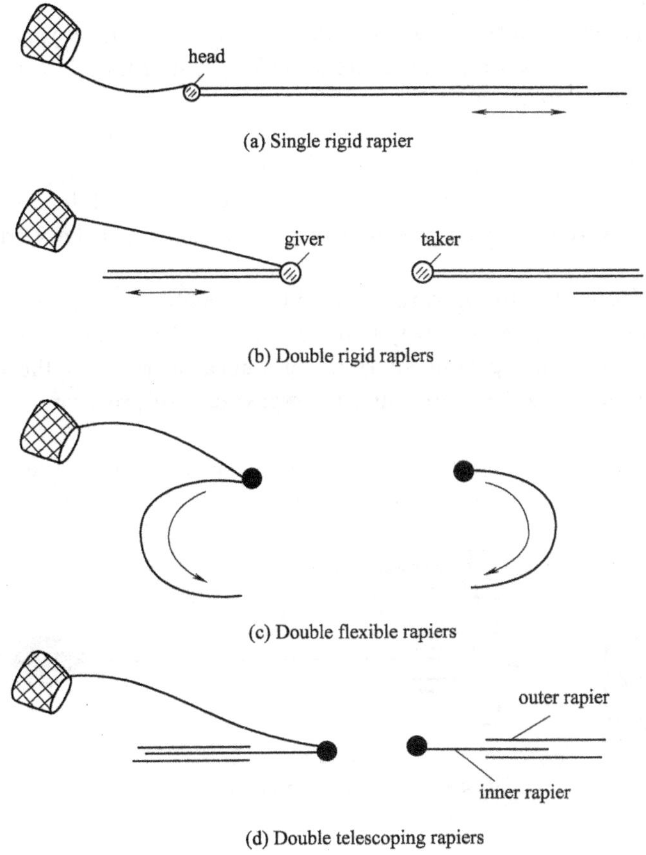

(a) Single rigid rapier

(b) Double rigid raplers

(c) Double flexible rapiers

(d) Double telescoping rapiers

Fig. 4 – 29 Rapier systems

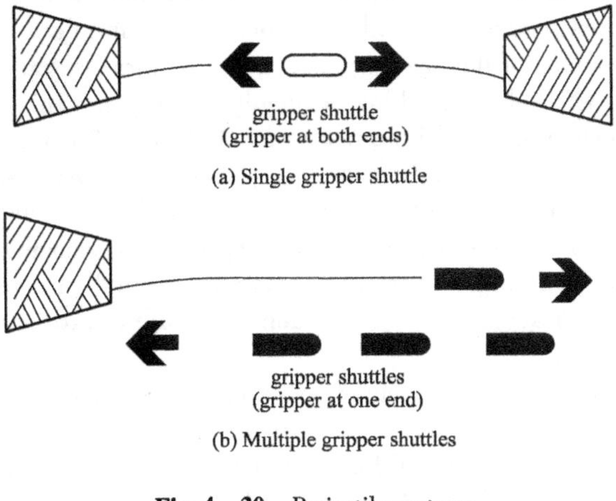

(a) Single gripper shuttle

(b) Multiple gripper shuttles

Fig. 4 – 30 Projectile systems

流体喷射织机 **4.2.2.3 Fluid Jet Looms**

Fluid jet looms employ a fluid, air or water, to carry the filling yarn through the shed. These looms allow high numbers of picks per unit time but have some limitations as to yarn type.

In an air jet loom, high pressure air carries the filling yarn through the shed. The major drawback to this type of loom is that the air diffuses quite rapidly so the width of the fabric is limited. Modern air jet looms use air guides or air guides with auxiliary jets to extend the distance that the yarn can be carried.

气流扩散

气流导纱器/辅助喷管

In a water jet loom, the filling yarn is carried by a high pressure jet of water. Since water diffuses much more slowly than air, wider fabrics can be made with water jet looms without using guides. However, because water is the carrier, there are restrictions as to yarn types and pre-weaving yarn processing, such as slashing.

织前

Fluid jet looms, both air and water jet, are schematically illustrated in Fig. 4-31.

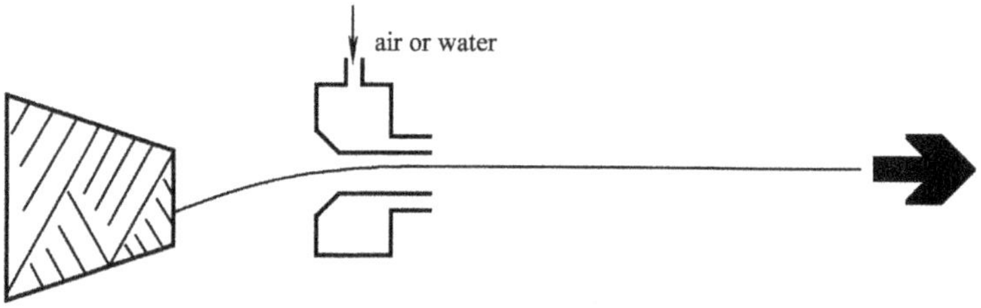

Fig. 4-31 Fluid jet systems

储纬系统 **4.2.2.4 Filling Accumulation Systems**

A distinction common to shuttleless looms is that the filling supply is exterior to the filling carrier. The obvious benefit in this system is the use of very large supply packages and thus a reduction in both extra preparation and down times.

To reap the benefits of this system for all types of yarn, a small price must be paid and that is the need for a filling accumulation system. This system removes yarn from the package prior to filling insertion. Without a filling accumulation system, due to the high speed of filling insertion, there is a good possibility that the filling yarn, instead of coming off the package smoothly, will tend to slough off. The accumulator reduces this hazard by removing the yarn smoothly one pick at a time and allowing the filling carrier to use this premeasured yarn instead of taking it directly from the package. Tension variation in the filling yarn is also reduced by using an accumulator.

塌脱

储纬器

There are two major filling accumulation systems illustrated in Fig. 4-32. The drum system winds a premeasured amount of yarn on a drum, in effect a small filling yarn package, and allows the filling carrier to take this yarn through the shed. The vacuum tube system collects the yarn in a U-shaped tube and the yarn is withdrawn from this tube during filling insertion.

真空管系统

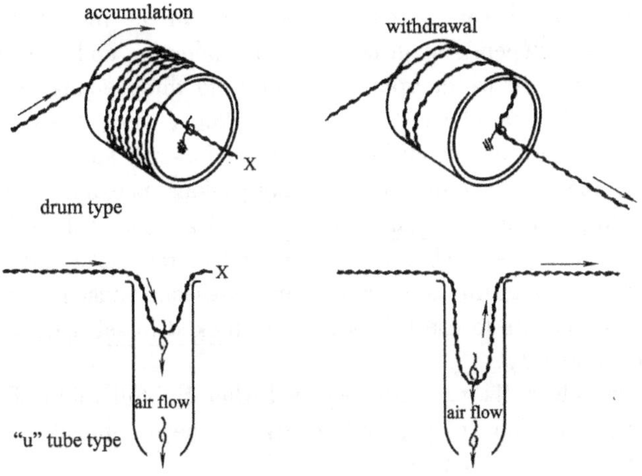

Fig. 4 – 32 Filling yarn accumulation systems

4.2.2.5 Modification of the Selvage 布边的改动

In yard goods, the outer edges are constructed so they will not ravel. These finished edges are called the selvages (self-edges) as shown in Fig. 4 – 33 (a) and 散开 布边

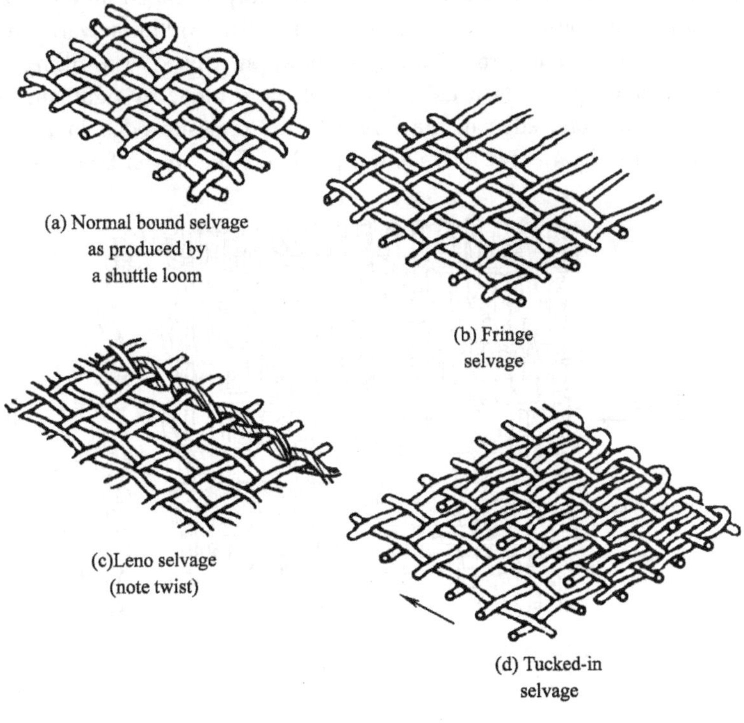

Fig. 4 – 33 Various forms of selvage

are often made more compact and stronger than the rest of fabric by using heavier and more closely spaced warp yarns than are used in the rest of the fabric. The kind of selvage used depends upon economy of production and expected use of the fabric. Selvages vary in width from one-quarter to three eights of an inch. The warp yarns always run parallel to the selvages. Proper use of the selvages can also prevent the bowing and bias conditions that occur in some fabrics. 卷边/斜边

In shuttleless looms, since the filling supply package is exterior to the carrier, it is not possible to produce two bound selvages. At least one of the selvages must be fringed as shown in Fig. 4 – 33(b). The fringed or raw selvage does not usually withstand the handling during subsequent processes and must be reinforced. The two common solutions are to use leno selvage or tucked-in selvage, as illustrated in Fig. 4 – 33(c) and (d). 光边 毛边 绞纱边/折入边

The tucked-in selvage is generally required when the full width of fabric is used in the end use application, e.g., bed sheets. In these applications the leno selvage is cosmetically unacceptable.

In the case of apparel fabrics, the selvages are generally not used, and therefore leno selvage may suffice. 服用织物

4.3 Woven Fabrics 机织物

A conventional woven fabric is a textile structure formed on a loom when two sets of yarns are interlaced at right angles in established sequences. The longitudinal yarns are known as the warp and one yarn of the warp is called an end. The warp is a sheet of parallel yarns held on the loom beam. The widthwise yarns are known as the filling or weft. One yarn of filling is known as a pick. The filling yarns are laid between two warp sheets usually one pick at a time, at right angles to the warp, and run from one edge of the fabric to the opposite edge in a continuous strand as shown in Fig. 4 – 34. The central portion of the woven fabric is called the 一根经纱 一根纬纱

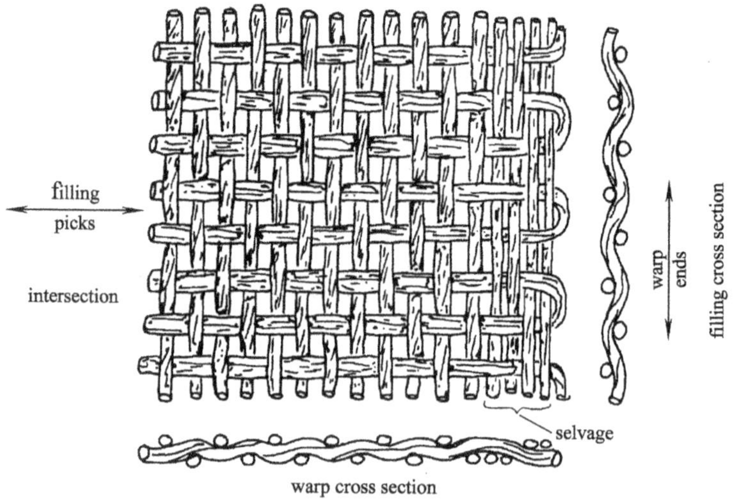

Fig. 4 – 34 Woven fabric

body and the edges are called the selvages. The selvages usually differ from the body of the fabric in both yarn and fabric construction.

布身

The fabric weave or design is the manner in which the warp and filling are interlaced. The pattern or repeat is the smallest unit of the weave which, when repeated, will produce the design required in the fabric. There are many ways of representing a weave, but the most commonly used method of representation is on square design paper.

织物组织
组织循环
意匠纸

On the design paper, the vertical columns of squares represent warp ends and the horizontal rows of squares represent filling picks. It is normal to use a filled-in square to indicate that the end is over the pick and a blank square to indicate that the pick is over the end.

Weave patterns can create varying degrees of durability in fabrics, adding to their usefulness and also to their appearance. Fabric weaves are limited only by the imagination of the fabric designer. However, there are three basic weaves in common use for the majority of fabrics. These structures are plain, twill and satin weaves with some variations. Important constructions are also obtained from the following weaves: pile, double, gauze, dobby, and Jacquard.

组织花纹
三原组织
平纹/斜纹/缎纹
起绒/双层/纹纱/小提花

4.3.1 Plain Weaves and Fabrics

平纹组织与织物

4.3.1.1 Plain Weave and its oaritons

平纹组织及其变化组织

(1) Plain weave

平纹组织

Plain weave, shown in Fig. 4 – 35, is the simplest of all weaves. One repeat of plain weave consists of two ends and two picks. Since each of the ends must weave differently, at least two harnesses are required, each weaving opposite the other. Despite their simplicity, plain fabrics form a high percentage of all woven fabric. Plain fabric has the highest number of interlacings as compared to other weaves causing it to be the firmest.

There are some fabrics which are considered derivatives of the plain weave. These are in effect, extensions of the simple interlacing. They also have this name because, theoretically, they can be produced on a loom with two harnesses moving in opposition. This harness movement is known as plain weave order.

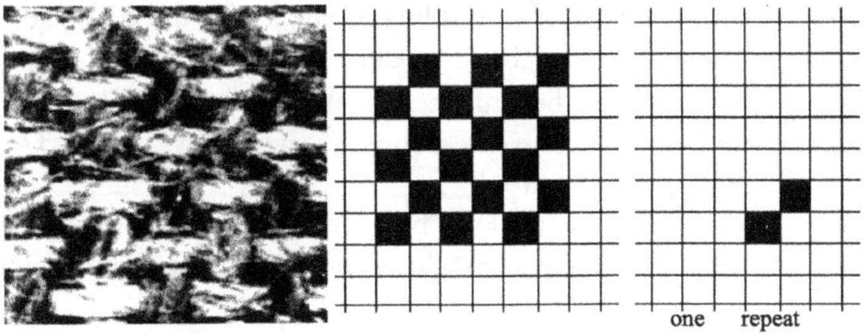

Fig. 4 – 35 Plain weave and cloth

经重平组织

(2) Warp Rib Weaves

In this, the extension of the plain weave is in the warp direction, as shown by Fig. 4 – 36(a), the warp weaves in the same order as in the plain fabric, namely, every two adjacent ends weave opposite to each other. The filling weaves in groups of 2, 3 or more and every group weaves opposite to the adjacent groups; the repeat is always on 2 ends but any number of picks may be involved. The weave is denoted by showing the number of picks in the group above and below a line, i. e. ,as $\frac{2}{2}$ warp rib weave or sometimes as $\frac{2}{2}$ warp rib weave. Although the weave is called warp rib it actually produces ribs in the filling direction. To obtain the best results it is usual to use fine warp and coarse filling in order to show the rib more clearly. The fabrics are normally lighter in weight than plain weave fabrics because of the lower level of crimp. The fabrics are also softer and more flexible than the plain fabrics.

变化经重平，花式经重平

Fancy warp rib weaves can be produced simply by changing the number of picks in the groups, which changes the width of the rib in the fabric as shown in Fig. 4 – 36(b).

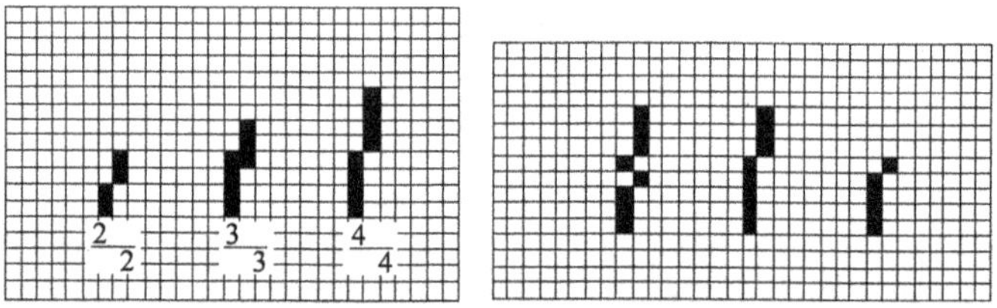

(a) Warp rib weaves (b) Fancy warp rib weaves

Fig. 4 – 36 Warp rib weaves

纬重平组织

(3) Filling Rib Weaves

In this case groups of ends are woven with each group in direct opposition to the adjacent groups. The repeat is always on 2 picks × any number of ends as shown in Fig. 4 – 37(a). The ribs produced in the fabric run in the direction of the warp. It is usual to use coarse warp and fine filling to emphasize the rib. The fabrics are normally stronger than the plain fabrics because of the low crimp level.

变化纬重平，花式纬重平

Fancy filling rib weaves can be produced in the manner already described, but by changing the number of ends in the group, as shown by Fig. 4 – 37(b).

方平组织

(3) Basket Weaves (Matt Weaves)

In this weave the extension is made in both directions so that groups of ends and picks are woven in the same way as single ends and picks are woven in the plain weave, The weave is denoted in the manner used for rib weaves, i. e. as $\frac{2}{2}$ basket or 2 × 2 basket. Fig. 4 – 38 (a) shows the designs for $\frac{2}{2}$ and $\frac{3}{3}$ basket

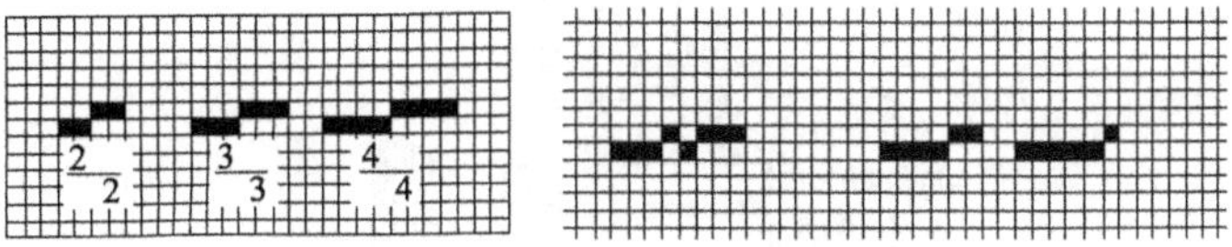

(a) Filling rib weaves (b) Fancy filling rib weaves

Fig. 4 – 37 Filling rib weaves

weaves. The numbers of ends and picks in the repeat are always equal to the addition of the two numbers denoting the weave. Thus a $\frac{2}{\ \ 2}$ basket repeats on 4 ends × 4 picks, a $\frac{3}{\ \ 3}$ basket repeats on 6 ends × 6 picks and so on. The weave is square and is mostly used with square constructions.

It is also possible to change the number of ends and picks in the groups to produce fancy basket weaves as shown in Fig. 4 – 38(b). 变化方平组织,花式方平组织

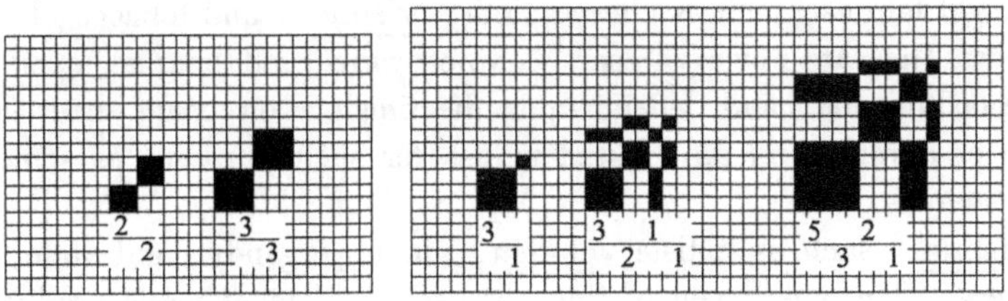

(a) Regular basket weaves (b) Fancy basket weaves

Fig. 4 – 38 Basket weaves

The construction and cross section figures for above three variations of plain weave are shown in Fig. 4 – 39.

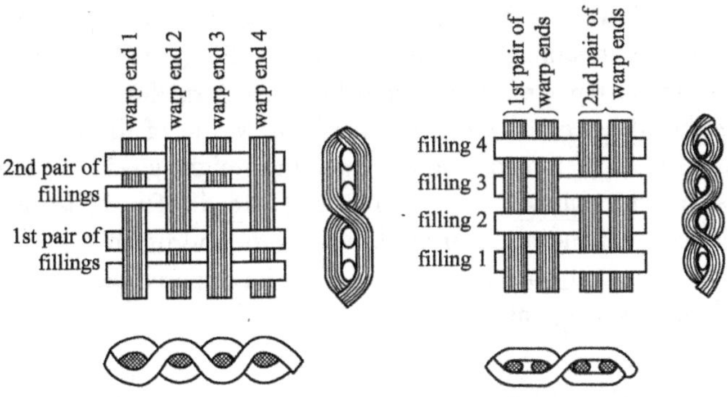

(a) $\frac{2}{\ \ 2}$ Warp rib (b) $\frac{2}{\ \ 2}$ Filling rib

Fig. 4 – 39

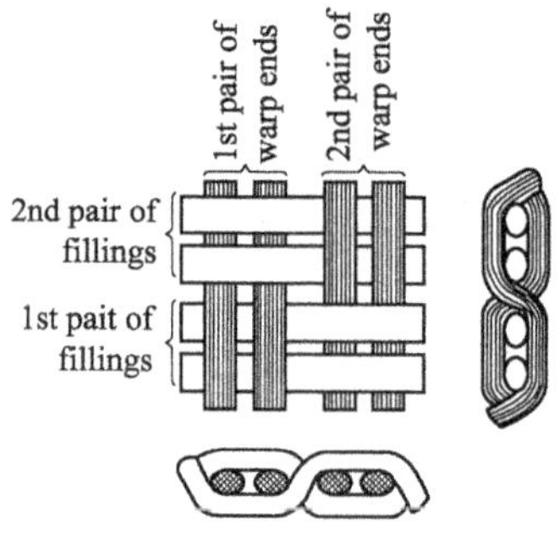

(c) $\frac{2}{2}$Basket

Fig. 4 – 39 Rib and basked weaves

平纹织物 4.3.1.2 Fabrics in Plain Weaves

轻薄/厚重

There is a wide variety of fabrics made of the plain weave constructed from every type of yarn composed of every kind of fiber. Plain woven fabrics range in weight and compactness from thin lightweights to compact heavyweights.

稀纱布/硬衬布 疏松结构

Lightweight fabrics, such as cheese cloth, crinoline, and tobacco cloth are low count sheers. They are woven of coarse, carded yarns producing an open construction. The texture and hand depend upon the finish. Low count sheer fabrics are not strong or durable and are limited in use for such purposes as screening and backing materials.

透明薄织物/雪纺/蝉翼纱/巴里纱 罩衫/衣服/衬衫

Sheer fabrics, such as batiste, chiffon, ninon, organdy, and voile, are high count sheers. They are woven of fine combed or filament yarns producing thin, transparent material. The texture and hand depend upon the kind of yarn, the yarn twist, and the finish. For their weight, high count sheers are reasonably durable and serviceable. They are used for such purposes as blouses, dresses, shirts, and for such home furnishings as curtains.

中厚织物

Medium-weight fabrics represent a very large group of materials. They are woven of medium size carded, combed, or filament yarns. These fabrics are given a variety of finishes and have a wide range of texture, hand, and appearance. They are serviceable; care of treatment depends upon the fiber, yarn, and finish used. They are used for all types of apparel and for home furnishings.

厚重织物 帆布/粗布/法兰绒/粗花呢 帷帐/室内装潢

Bottom-weight fabrics are the heavier materials woven of thicker yarns but have about the same range of thread count as medium-weight goods. These fabrics include canvas, crash, flannel, homespun, and tweed. The construction may be very compact and, depending upon the kind of yarn and finish, can be quite durable. Bottom-weight plain woven fabrics are used for apparel and for such home furnishings as drapery and upholstery.

斜纹组织与织物 4.3.2 Twill Weaves and Fabrics

斜纹组织 4.3.2.1 Twill Weaves

对角线

Twill is the second basic weave and is characterized by diagonal lines running at angles to the fabric. A twill weave is denoted by using numbers above and below a line (such as $\frac{2}{1}$ twill). These numbers represent the position of the first end throughout the number of picks in the weave. For simple twill weaves, the repeat consists of as many ends and picks as the sum of the numbers above and below the line. For example, the above twill repeats on three ends and three picks (2 + 1 = 3). The position of the first end is over the first and second pick (denoted by 2 above the line) and under the third pick (denoted by 1 below the line). The positions of the remaining ends can then be determined by repeating the motion of the first end but beginning the motion on successive picks. Since each of the three ends weave differently, the above fabric requires a minimum of three harnesses to produce.

The numbers above and below the line also give a clue as to which yarn, the warp or the filling, will be seen more on the face of the fabric. If the number above the line is the greater, the warp yarn will predominate and the design is called warp face twill. If the number below the line is the greater, the filling yarn will predominate and the design is called filling face twill. If both numbers are the same, the twill is said to be balanced. These twills are shown in Fig. 4 – 40.

经面斜纹
纬面斜纹

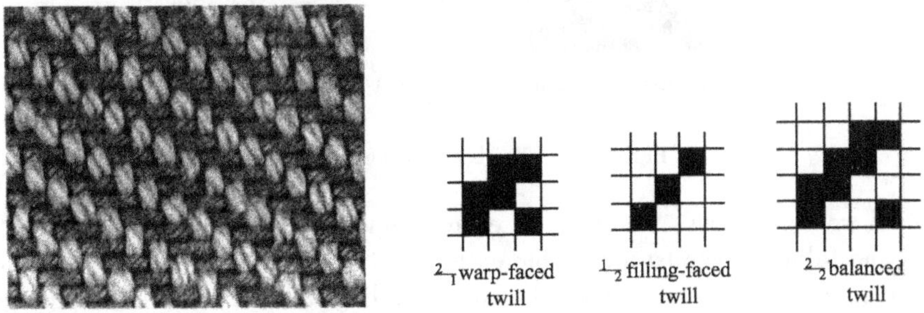

Fig. 4 – 40 Twill weaves and cloth

In the twill weave, the filling yarn interlaces more than one warp yarn but never more than four, as strength would be sacrificed by so doing. On each successive line, or pick, the filling yarn moves the design one step to the right or to the left, thus forming the diagonal. What ever the direction of the diagonal on the face of the fabric, the design runs in the opposite direction on the reverse side.

Twill weaves are named according to the number of harnesses required to make the design. A 3-harness or 4-harness twill is frequently used. The word "shaft" may be substituted for "harness", as in 3-shaft or 4-shaft twill.

四页斜纹，四枚斜纹/综框

The twill weave is always given a direction such as right-hand if the twill line runs from lower left to upper right or left-hand if the twill line runs from lower right to upper left. The right-hand and left-hand versions of the same twill are shown in Fig. 4 – 41 and Fig. 4 – 42. Although there is no advantage of one over the other, the direction of the diagonal can aid in the recognition of the face of the fabric.

Fig. 4 – 41 A 3-shaft, right-hand, filling-faced twill

(a) Even $\frac{2}{2}$, right-hand (b) Uneven $\frac{1}{2}$, left-hand

Fig. 4 – 42 The basic diagonal of twill

Twill weaves are also classified as even or uneven according to the number of warp and filling yarns that are visible on the face of the fabric. The even twill, for example, shows an equal number of warp and filling yarns in the recurring design, such as two over and two under. This pattern makes what is called a 4-shaft twill, and it requires 4 harnesses. An uneven four shaft twill, for example, that has one warp yarn riding over three filling yarns is referred to as a three up and one down, or $\frac{3}{1}$. On the other hand, a 3-shaft twill that has one warp yarn riding under two filling yarns is referred to as a one up and two down, or $\frac{1}{2}$ (see Fig. 4 – 42).

Most twill weaves are uneven. An uneven twill may show more warp than filling yarns in the recurring design. If or more filling yarns than warp yarns show on the face. Warp-face twills are generally stronger than filling-face twills because the stronger warp yarns on the face of the fabric can take more abrasion and wear. Warp-face twills generally have much more warp than filling yarns, consequently, such fabrics hold their shape better and drape better due to the warp's greater twist and resilience.

Many combinations and variations of twill constructions are possible. These produce interesting pattern effects, which can be further enhanced by yarns of different colors and textures (Fig. 4 – 43).

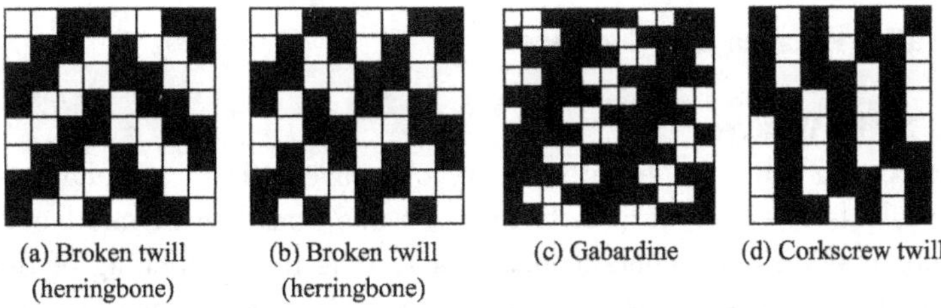

Fig. 4 – 43 The variations of the twill weave

4.3.2.2 Fabrics in Twill Weaves

斜纹织物

Changes in the direction of the twill diagonal lines produce variations, such as the herringbone, steep twill, corkscrew, entwining, and fancy twills (Fig. 4 – 44). Increased ornamentation may be obtained by varying the slant of the diagonal and yarn colors. The values of the twill weave include its strength and drapability. The diagonally arranged interlacings of the warp and filling provide greater pliability and resilience than the plain weave. Also, twill fabrics are frequently more tightly woven and will not get dirty as quickly as the plain weave, though twills are more difficult to clean when they do get soiled. The yarns are usually, closely beaten, making an especially durable fabric. Twill weaves are therefore commonly used in men's suit and coat fabrics and for workclothes, where strong construction is essential.

人字斜纹/急斜纹/芦席斜纹/装饰效果/倾斜度

柔韧性

男士西装/外套织物/工作服

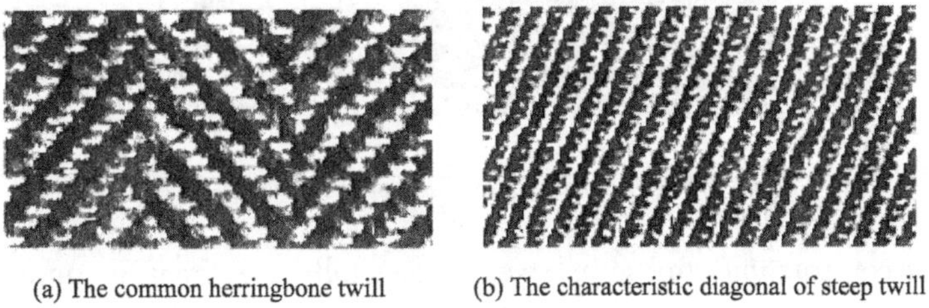

Fig. 4 – 44 Twill weave fabrics

The steepness of the diagonal can indicate strength and durability in the fabric. In order to obtain a steep twill, more warp yarns must be used than filling yarns. And since warp yarns have a higher twist and are stronger than filling yarns, the steeper the twill the stronger the fabric is likely to be.

Although there are lightweight twills, the majority of fabrics in twill weave are medium weight, and there is a substantial variety of bottom-weights. Depending upon fiber, yarn, construction, and finish, they are used for a wide range of apparel, such as dresses, suits and coats, and home furnishings, such as drapery and upholstery.

Cloth made of staple yarns are usually left-hand twills. Wool-type fabrics are usually right-hand twills. Fabrics made of filament yarns are also right-hand twills.

缎纹组织与织物 4.3.3 Satin Weaves and Fabrics

缎纹组织 4.3.3.1 Satin Weaves

Satins make up the third basic category of weaves. They are characterized by having long floats of either warp or filling yarn and nonadjacent interlacing points. As many other weaves can contain long floats, it is the positioning of the interlacing points which makes the satin weave unique. This positioning is accomplished 飞数 through the use of a counter. The satin weave may have a warp-face or filling-face 经面缎纹 construction. Fig. 4 – 45 shows a warp satin in which the floats are made up of 纬面缎纹 warp yarns and a filling satin, or sateen, where the floats are caused by the filling.

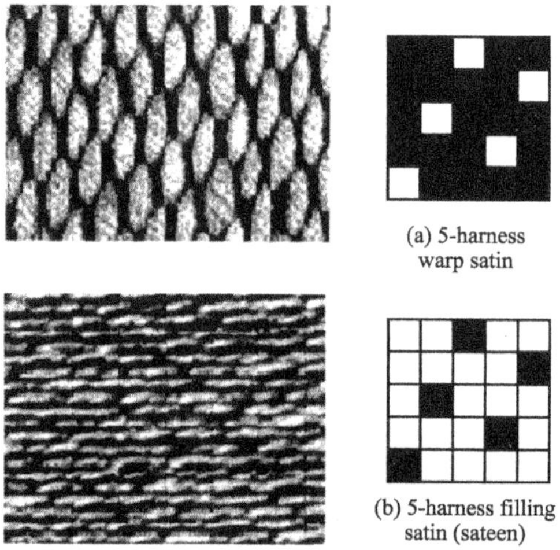

Fig. 4 – 45 Satin weaves and clothes

In basic construction, the satin weave is generally uses from five to as many as twelve harnesses, producing a 5-shaft to 12-shaft construction. It differs in appearance from the twill weave because the diagonal of the satin weave is not visible; it is purposely interrupted in order to contribute to the flat, smooth, lustrous surface desired. There is no visible design on the face of the fabric because the yarns that are to be thrown to the surface are greater in number and finer in count than the yarns that form the reverse of the fabric.

Warp-face satin is woven so that the warp may be seen on the surface of the fabric. Since the warp lies on the surface and interlaces only one filling at a time, the 浮长 lengths of warp between the filling are called floats. These floats lie compactly on the surface with very little interruption from the yarns going at right angles to them. Reflection of light on the floats gives satin fabric its primary characteristic of luster, which appears in the direction of the warp (see Fig. 4 – 46).

The filling-face satin weave is also called the sateen weave. In this construction, the filling yarn lies on the surface of the fabric as it passes regularly over and under the warp yarns. The floats are consequently made up of the filling yarns, and the luster appears in the filling direction (see Fig. 4 -47).

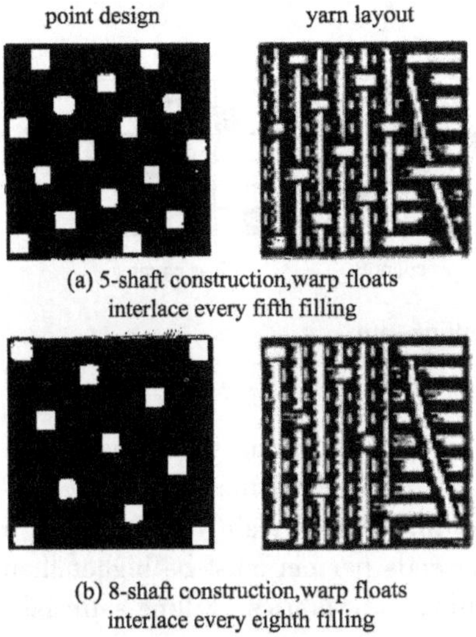

Fig. 4 -46 Warp-face satin weaves

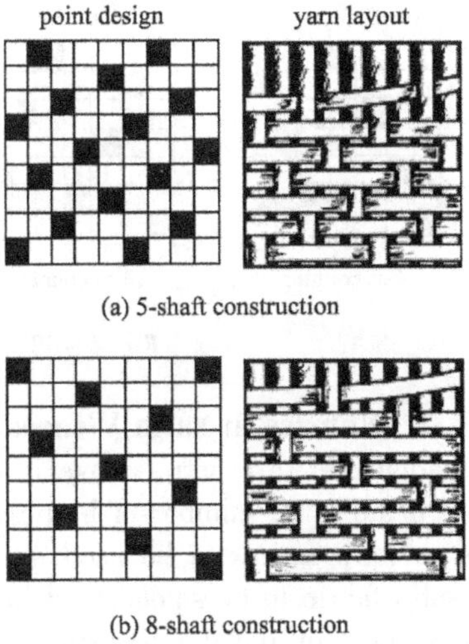

Fig. 4 -47 Filling-face satin (sateen) weaves

Satin weaves are designated by the number of harnesses needed to produce them and by which yarns make up the floats. They repeat on the same number of ends and picks as there are harnesses. For example, a 5-harness sateen repeats on 5 ends and 5 picks requires 5 harnesses to produce and has floats made up of filling yarns.

For every number of ends and picks in the repeat there is more than one arrangement of interlacing points. Finding all the possible counters for a satin weave is simple if the following rules are observed:

① For any number of harnesses greater than or equal to 5, find all pairs of numbers which sum to the number of harnesses.

② Eliminate all pairs containing:

(a) The number 1.

(b) Numbers with a common divisor other than 1. 公约数

Any one of the numbers in the remaining pairs may be used for the counter as long as only one is used for the entire repeat. Mixing of counters within a repeat is not allowed.

After laying out the repeat size, to determine the position of the interlacing points once the counter is known is a relatively simple task. The first interlacing point is usually taken as the lower left-hand square. The next interlacing point is found by moving to the right the same number of squares as the desired counter. The interlacing point is the square on the next pick directly above this square. The next interlacing point is found by repeating this procedure using the last interlacing point as the new starting point. Once all the interlacing points are found, the weave may be developed by marking only the interlacing points in the case of a filling satin or, in the case of a warp satin, marking all but the interlacing points. All possible arrangements of a 7-harness filling satin are shown in Fig. 4-48.

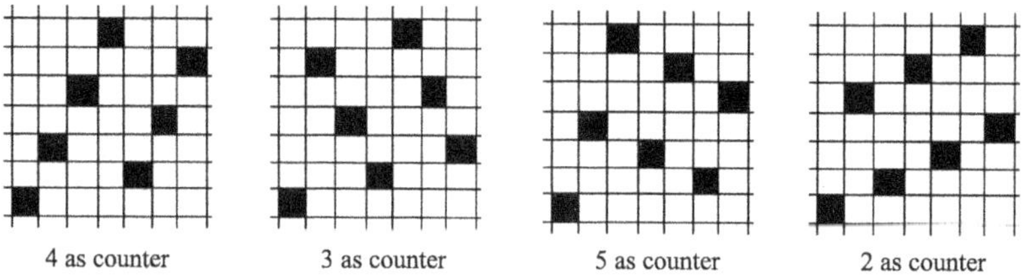

Fig. 4-48 7-harness filling satin

缎纹织物

4.3.3.2 Fabrics in Satin Weaves

Although the warp satin weave is the back side of the filling satin weave produced on the same number of harnesses, it is not true to say that the fabric produced as warp satin can be used as a filling satin. This is mainly because for a warp satin fabric to be smooth and lustrous, the ends per cm must be higher than the picks per cm and the opposite is true for filling satin fabrics. Filling satin fabrics are normally softer and more lustrous than warp satins, but the warp satins are usually stronger than filling satins. In both fabrics, if a heavy construction of filament yarns is used, the fabric tends to be stiff and does not drape easily. However, the length of the float can balance the effect of the construction, but it must be realized that longer floats have disadvantages in affecting the fabric serviceability. Long float satins are useful in Jacquard designing or in combination with other weaves but are rarely used elsewhere.

时髦服装 / 衬里

The long floats found in the satin weave might be considered a disadvantage because they represent a minimum of interlacings, and therefore a potential weakness in the fabric. Furthermore, to increase the smoothness and luster of the fabric, the yarns are given a minimum of twist and are therefore relatively weak. The longer the float, the greater the chances that the surface of the fabric will snag, roughen, and show signs of wear. However, the luster makes the fabric suitable for dressy wear and the smoothness, for use as lining.

身骨/孔隙 / 晚礼服

Satin fabrics drape well because the weave is heavier than the twill weave, which, in turn, is heavier than the plain weave. More harnesses are used for satin weave, thus compressing a greater amount of fine yarn into a given space of cloth. This compactness gives the fabric more body as well as less porosity, which makes the fabric warmer. The quality of drapability also makes satin fabrics preferable for evening wear, and the warmth contributes to its value as lining material.

Most fabrics made with the satin weave are medium weight; a lesser number are heavier weights. Both staple fiber and filament yarns are used.

锦缎 / 枕套及褥罩织物/威尼斯缎纹织物

The staple fibers most frequently used are cotton and rayon for such fabrics as damask, sateen, ticking, and Venetian cloth for use in apparel and home furnishings.

Linen is used for damask home furnishings, principally table cloths.

Silk, rayon, acetate, and nylon filament yarns are usually used for such material as brocade, damask, merveilleux, and satin. 锦缎/斜纹背高级里子缎

Filling-face satin fabrics are generally of medium-weight and made of staple fiber yarns. They are usually made of cotton or rayon into damask, sateen, and ticking. When linen is used, the cloth is likely to be a damask.

4.3.4 Drawing-in Draft, Lifting and Reed Plans

穿综图、提综图与穿筘图

The design (D) represents the manner in which the interlacing between the warp and filling yarns takes place. The drawing-in draft (DID) shows the arrangement of the warp yarns in the different harness frames. The lifting plan (LP) or chain plan (CP) represents the pattern in which the harness frames are lifted or lowered for every pick in the repeat. The reed plan (RP) shows the arrangement of the warp yarns in the reed dents.

组织图 穿综图 提综图/纹板图 穿筘图

Normally 2 ends per dent for the fabric body and 4 ends perdent for the selvage is a reasonable combination. However, in many cases, 3 or 4 ends/dent are used. The reed plan can be either regular or irregular depending on whether or not the same number of ends per dent is used regularly across the width of the body of the warp. Some designs require the use of different numbers of ends per dent in the fabric body to produce certain effects in the fabric.

Normally, the drawing-in draft is placed either above or below the design with a space in between them. It can be shown by filling in the squares or by using crosses. The lifting plan is usually placed to the right-hand side of the design and it is always based on the filling in of the squares; again, a space between the design and the lifting plan is necessary. The reed plan is normally placed below the design and can be indicated either by filling in squares or by using brackets.

Fig. 4-49 shows the projections of design, drawing-in draft, lifting plan and reed plan for an 8-end sateen weave. In this case the draft has to be straight, because every end in the repeat is woven differently, and there are 8-harnesses. The

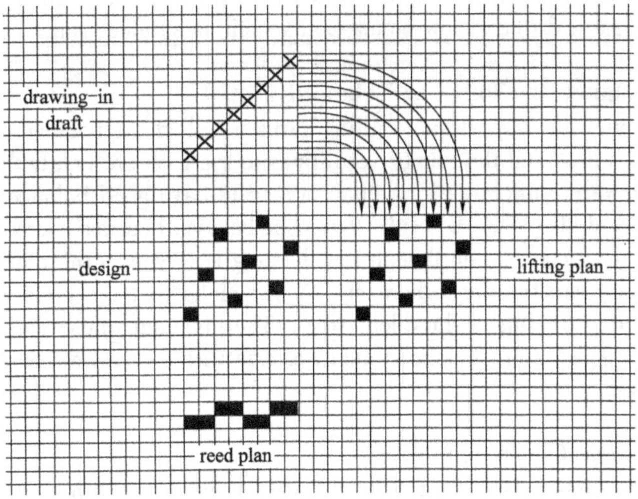

Fig. 4-49 Drawing-in draft, lifting and reed plans

筘号 ends are dented 2 ends per dent. The reed number(number of dents/inch or dents/cm) and the end density determine the width in the reed (WIR) of the warp. Should the drawing-in draft be straight, then the design and the lifting plan are similar, but if the drawing-in draft be in any other form, e. g., skip draw, then the lifting plan will be dissimilar to the design. If any two of the projections—design, drawing-in draft, and lifting plan—are given, the third can be deduced from them.

织物基本术语 4.3.5 General Terms Related to Fabrics

织物正反面 4.3.5.1 Back and Face of a Fabric

正面 One side of fabric is referred to as the face or the right while the opposite side is 反面 referred to as the back or the wrong. The face or the right is that side of the fabric which presents a better appearance. This is usually the side of the fabric intended to be seen when the fabric is in use.

织物经、纬向 4.3.5.2 Warp and Filling of a Fabric

The length of the fabric indicates the warp yarns. Also, if a piece of the fabric shows part of the selvage or finished edges, which is the firm edge of the cloth, then the yarns parallel to the selvage are warp yarns. The opposite yarns are the filling yarns..

织物密度 4.3.5.3 Fabric Count (or Count of Cloth)

The number of ends and picks per cm(or inch) in a fabric, expressed as $E \times P$, is known as fabric count or count of cloth. The count of cloth is deferent from the count of yarn which indicates the weight and diameter of the yarn.

If the count of the cloth is 80 warps (ends) and 80 fillings (picks) to the inch, the count is expressed as 80 ×80, or 80 square. If there are 60 warps and 50 fill-高密织物 ings to the inch, the count is expressed as 60 ×50. There are, then, higher-count 直贡缎/低密织物/手术纱布 cloth such as satin drill and low-count cloths such as surgical gauzs.

织物平衡(平衡织物) 4.3.5.4 Balance of Cloth (or Balanced Cloth)

The proportion of warp yarns to filling yarns is called the balance of a cloth. If the number of warps and the number of fillings to the inch are nearly the same (not 被单布 more than ten yarns difference), a cloth is said to have good balance. A sheeting with 61 ends and 40 picks (61 ×40) has poor balance because there are too many ends and too few picks. Even though the sheeting is woven in the plain weave, or-滑移 dinarily a strong construction, there are so few picks that the ends will slip over 纰裂效应 them very easily, causing a shredded effect. The gingham whose count is 60 ×50 would be considered a fair balanced cloth. Gauze with a count of 28 ×24 also has a good balance.

织物结构,织物规格 4.3.5.5 Fabric Construction

Fabric construction is indicated by the yarn number in both warp and filling, fabric count, the fabric width and the fabric weight. This is written in the following form:

width, $g/m^2(oz/yd^2)$, $E \times P$, $Tt_w \times Tt_j$

4.3.5.6 Warp and Weft Crimp in Fabric

织物经纬纱屈曲

The interlacing causes the warp and weft yarns to assume a wavy configuration, in general terms, this waviness is called crimp. It has important effects on the dimensions and performance of the fabric; for example, the existence of crimp means that the lengths of filling or warp yarn required are greater than the width or length of the cloth. Since the basis weight of the fabric depends not only upon the linear density of the yarns used but also on the total length of the yarns assembled in the fabric, it is apparent that the crimp has to be taken into account. 屈曲

The percentage of change in length of yarn woven into fabric as compared to the fabric length or width corresponding to warp crimp or filling crimp respectively. If L represents the yarn length before weaving and X the yarn length in the woven fabric then the crimp, R, is given by

$$R = \frac{L - X}{X}$$

The percentage change in length of yarn woven into fabric compared to the original yarn length is known as the contraction. Using the same representation as for crimp, the contraction, T, is given by 缩率

$$T = \frac{L - X}{L}$$

4.3.5.7 Fabric Cover Factor

织物覆盖率

Fabric cover factor is the percent area projected by warp and filling yarns in a given fabric area. The diagram in Fig. 4-50 shows in a qualitative manner what is meant by cover factor which is concerned only with the geometry.

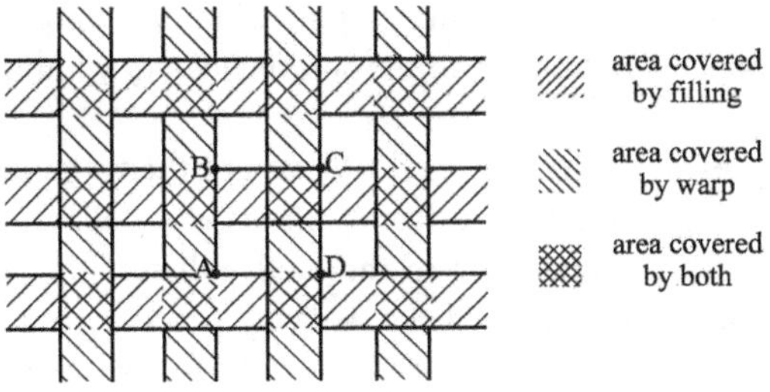

Fig. 4-50 Illustration of cover factor

If in a plain woven fabric, the same yarn number is used for warp and filling and the fabric has the same ends and pick per inch (i.e., square fabric) and if the construction of the fabric is such as to leave a space between the yarns equal to the yarn diameter, the warp cover factor will be 50% and the filling cover factor will also be 50%.

The percentage fabric cover factor in this case is 75%. Note the fabric cover is not the sum of the warp and filling covers, because the overlapping areas of yarns should not be counted twice.

A fabric of open construction is one with a cover factor 25% to 50%, the regular construction with 51% to 75%, the close construction with 76% to 100%.

If any of the warp or filling cover factor is 100%, the fabric cover will be 100%.

织物质地 4.3.5.8 Fabric Texture

Texture refers to the surface characteristic of a fabric; i. e., smooth or rough, and the spacing of ends and picks in the fabric; i. e., loose or tight.

The yarns used for warp and filling are frequently not of the same diameter. Usually, there are more warp yarns than filling yarns to the centimeters, because the strain on a fabric that is being used comes primarily on the warp. Some cloths, like ginghams (方格色织布), are closely woven; others, like voile, are loosely woven. If the cloth is held to the light, the porosity of the fabric or the closeness of the weave can be discerned. Ordinarily, a closely woven fabric keeps its shape better, shrinks less, slips less at the seams (缝线) and wears longer than does a loosely woven cloth of similar texture and weight.

织物重量 4.3.5.9 Fabric Weight

Generally, the weight of a piece of fabric (which is usually measured in g/m^2) is the combined weight of the warp and the filling yarns, but if the fabric is in the loom state, allowance (余量) must be made for the weight of size material on the warp. Also allowance has to be made for the crimp which changes in finishing.

思考题 Review Questions

1. Why the yarn preparation is necessary? What processes are needed?
2. Differentiate the cross-wound package from parallel-wound package.
3. What are the purposes of winding, quill winding, warping and slashing?
4. How about your understanding to tension devices in winding and warping processing?
5. How many types of creels are used? What are their characteristics?
6. What is weaving? What is a loom? Prepare a talk on the warp passage through the loom, from the back rest to the cloth roller.
7. Describe the basic operations of the weaving process, and the names for specific parts of the loom. Can weaving be done without a shuttle? How?
8. Compare the three main shedding mechanisms, the cam, the dobby and the Jacquard.
9. Why shuttle looms are still widely used around the world? Why the shuttle looms will be replaced by the shuttleless looms?
10. Describe the types of shuttleless looms you know.
11. What is the selvage in a shuttle loom and selvage modification in a shuttleless loom? Which is the best? Why?
12. Make a definition for following terms: fabric weave, yarn float, point-paper design, intersection, warp point, filling point, end, pick, weave repeat, warp-

faced weave, filling-faced weave.

13. Differentiate the characteristics among the plain, twill, satin or sateen weave.

14. Draw a point-paper design to illustrate the following weaves: the plain weave; all 4-shaft twills; possible 5-shaft satin weaves ; possible 8-shaft sateen weaves.

15. Compare the advantages and disadvantages of a cloth made in the plain, twill and satin or sateen weaves. Is it of any advantage to the consumer to know the basic differences among the three foundation weaves? Why?

16. Which of the three basic weaves is:

(1) The most durable?

(2) The most beautiful?

(3) The most inexpensive to produce?

(4) The most serviceable? Give reasons for each answer.

17. Indicate how to make the projection of weave design, drawing-in draft, lifting plan, and the reed plan for a certain weave on a loom.

18. Give the answer for:

(1) How does count of cloth differ from count of yarn?

(2) What is meant by balance of a cloth?

(3) Why is the count of cloth important to a buyer of cotton or linen fabrics?

19. Why do you think the crimp and cover factor are an important elements of fabric? How to calculate it?

Suggested Activities 实训题

1. Use strips of paper about 1/4 inch wide. Two colors are preferable—one for warp and one for filling. Interlace these strips of paper to form a paper mat of plain weave. Then make the twill and satin weaves. These mats should be kept in the textile notebook or manual.

2. Under the teacher's help do the following experiments:

(1) Identification of warp and filling.

(2) Identification of weave. Mount each sample on a sheet of paper with the warp running lengthwise of the paper. With the aid of a pin or a pick glass, look at the filling and count the number of warps, it goes over and under. Draw a point-paper design to illustrate the weave. 织物分析镜

(3) Count of cloth. Unravel a number of yarns both ways to make wide, frayed edges. If the fabric is of a light color, put it against something black, and vice versa. With the aid of a magnifying glass, count the number of yarns to the 1/4 inch, first one way of the cloth and then the other way. Express the count of the cloth by giving the number of picks and ends to the inch. Does the cloth have good balance? Why? Will the cloth wear well? Why?

(4) Test for durability of the weave. Grip opposite edges of the cloth tightly. Put your thumbs together and press down hard on the fabric. Does the cloth tear? Does the weave become badly distorted? Is the weave durable?

Chapter 5 Knitting and Knit Fabrics
针织与针织物

Knitting is the second mostly widely used method of constructing fabrics, and defined to be the formation of fabric by the intermeshing of loops of yarn. Unlike weaving, which requires two yarn sets, knitting is possible using only a single set of yarns. The set may consist of a single yarn or a single group of yarns.

相互圈套

The origins of knitting are rather obscure but some of the earliest examples of true knitting are stockings from Egypt which are dated at around 1500 - 1200 B. C.. However, it was not until 1589 that the process was mechanized, when the Reverend William Lee of Calverton in Nottingham in the UK was credited with the invention of the knitting frame. This machine was a flat-bed unit and used spring needles. It remained substantially the same for over 200 years. Since that time there have been many improvements in knitting, but the principle of interlooping yarns remains basic to the process.

长筒袜

钩针

Knitting yarn types include flat and textured filament, spun, blends of natural and man-made fibers, and new developed fibers such as Coolmax, PBT, PTT, Modal, Amicor, bamboo fiber and soybean fiber, etc.. Furthermore, existing knitting machines have been successfully adapted to use various types of high-performance fibers, including glass, carbon, aramid and even ceramics.

The usage of knitted fabrics ranges from hosiery, underwear, sweaters, slacks, suits, and coats, to rugs and other home furnishings. Moreover, knitted fabrics have the potential of being used in engineering structures. And it can be expected that knits will continue to be popular and to be found in a wide variety of applications in the years to come.

针织基本术语 5.1 General Terms of Knitting

针织分类 5.1.1 Sort of Knitting

When knitting is done by hand, the loops are held on one knitting needle and the yarn is pulled through each loop individually to form the new row of stitches. This row of stitches is subsequently connected to the next row. In machine knitting there is a needle for every stitch and it is the action of this needle that pulls the yarn through the old loop to form the new stitch.

织针

线圈

Machine knitting consists of forming loops of yarn with the aid of thin, pointed needles or shafts. As new loops are formed, they are drawn through those previously shaped. This interlooping and the continued formation of new loops produce knit fabrics. Two general methods are used: weft or filling knitting and warp knitting. 纬编/经编

Weft knitting is characterized by loops forming through the feeding of the weft yarn at right angles to the direction in which the fabric is produced [Fig. 5 - 1 (a)]. Warp knitting, on the other hand, is characterized by loops forming through the feeding of the warp yarns, usually from warp beams, parallel to the direction in which the fabric is produced [Fig. 5 - 1(b)]. More precisely, warp knitting is affected by interlooping each yarn into adjacent columns of wales as knitting progresses. 纵行

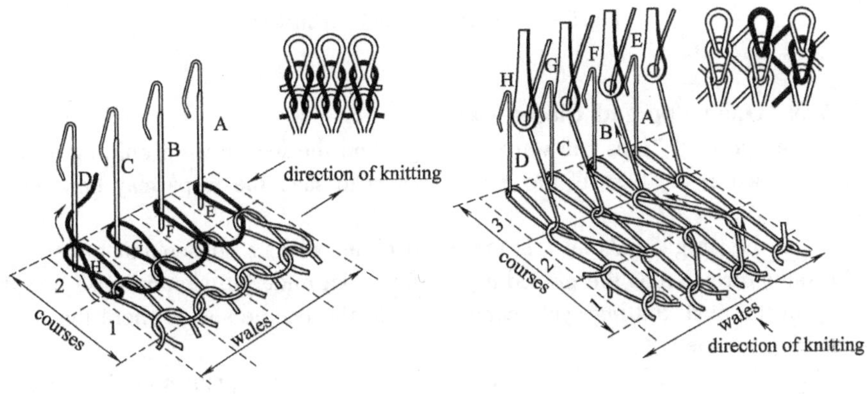

(a) Weft knit (b) Warp knit

Fig. 5 - 1 Schematic of knitted fabrics

5.1.2 Knitted Loop 针织线圈

Elements of knitted loop structure are introduced as following. As for a weft knitted loop, it consists of a needle loop and a sinker loop, while for a warp knitted loop, it consists of a needle loop and a underlap loop. 圈干/沉降弧 延展线

5.1.2.1 Needle Loop 圈干

The needle loop ($H + L$ in Fig. 5 - 2) is the basic unit of knitted structure. When tension in the fabric is balanced and there is sufficient take-away tension 牵拉张力 during knitting, it is an upright noose formed in the needle hook. It consists of a head (marked by H) and two slide limbs or legs (marked by L). At the base of 针编弧/圈柱 each leg is a foot (marked by F), which meshes through the head of the loop formed at the previous knitting cycle, usually by that needle. The yarn passes from the foot of one loop into the foot and leg of the next loop formed by it.

5.1.2.2 Sinker Loop 沉降弧

The sinker loop (marked by S in Fig. 5 - 2) is the piece of yarn that joins one weft knitted needle loop to the next. Sinker loops show on the opposite side of the fabric to the needle loops because the needle loop is drawn onto the opposite side from which the yarn was originally fed.

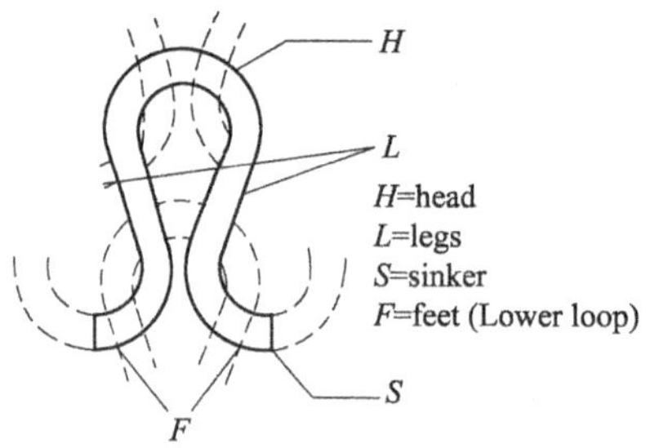

Fig. 5 - 2 Single loop and terminology

开口线圈与闭口线圈

5.1.2.3 Open Lap and Closed Lap

Loops are termed "laps" in warp knitting. And the feet in warp knitting may be open or closed at the base of the loop. That is to say, the loops may be open or closed.

开口线圈/针后垫纱 针前垫纱

An open lap [Fig. 5 - 3(a)] is produced either when a subsequent underlap is in the same direction as the preceding overlap or an underlap is omitted so that the overlap of the next knitting cycle commences in the needle space where the previous overlap finished.

闭口线圈

A closed lap [Fig. 5 - 3(b)] is produced when a subsequent underlap shogs in the opposite direction to the preceding overlap, thus lapping the same yarn around the back as well as around the front of the needle.

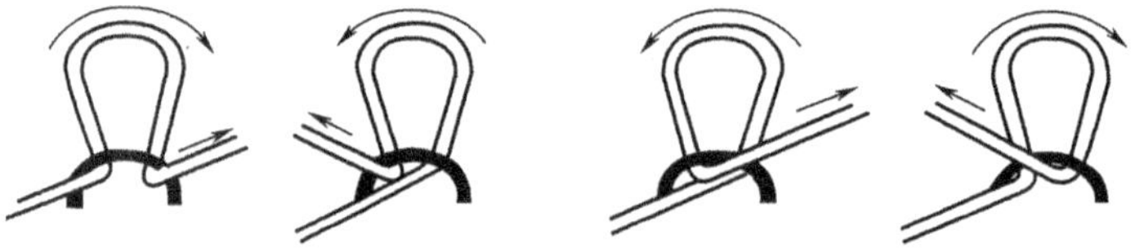

(a) Open lap (b) Closed lap

Fig. 5 - 3 Warp knitted laps

正面线圈与反面线圈

5.1.2.4 Face Loop and Reverse Loop

The face side of the stitch [Fig. 5 - 4(a)] shows the new loop coming towards the viewer as it passes over and covers the head of the old loop. It is referred to as

正面

the right side. Face loop stitches tend to show the side limbs of the needle loops or overlaps as a series of interfitting "V"s. The face loop-side is the underside of the stitch on the needle.

The opposite side of the stitch to the face loop-side is illustrated in Fig. 5 - 4 (b). It shows the new loop meshing away from the viewer as it passes under the

反面/反面线圈

head of the old loop. It is referred to as the left side. Reverse stitches show the sinker loops in weft knitting and the underlaps in warp knitting most prominently on the surface. The reverse loop side is the nearest to the head of the needle because the needle draws the new loop downwards through the old loop.

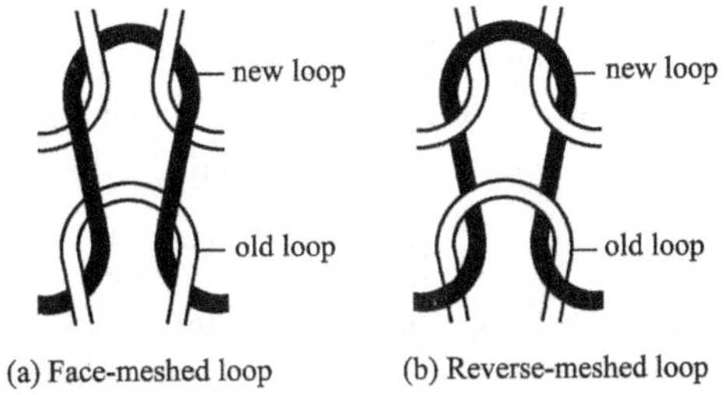

Fig. 5 –4 Face loop and reverse loop

5.1.2.5 Courses and Wales

线圈横列与线圈纵行

In the construction and analysis of knits, two terms are used frequently: course and wale (Fig. 5 –5).

横列

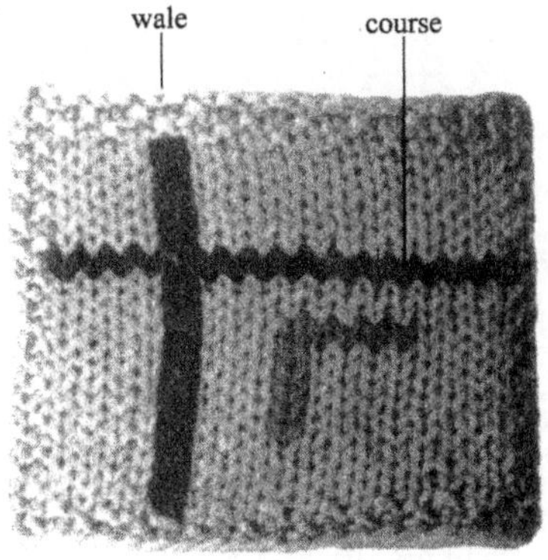

Fig. 5 –5 Weft knit. Dark yarns illustrate the wale and course

A course is a predominantly horizontal row of needle loops (in an upright fabric as knitted) produced by adjacent needle during the same knitting cycle.

A wale is a predominantly vertical column of intermeshed needle loops generally produced by the same needle knitting at successive (not necessarily all) knitting cycles. Wales are connected together across the width of the fabric by sinker loops or underlaps. Wales show most clearly on the technical face and courses on the technical back of single needle bed fabric.

工艺正面

工艺反面/单针床

5.1.3 Knitting Elements

针织机件

The actual knitting elements include the needles, sinkers, the needle bed or frame that holds the needles, and the yarn carriers that lay the yarn in knitting position. Here two main knitting elements are introduced — the needles and the sinkers.

织针 **5.1.3.1 Needles**

In both warp and weft knitting, the principal mechanical elements used to form loops are needles. In modern knitting, the three major types of needles are the latch needle, spring-bearded needle and compound needle.

舌针 (1) Latch Needle

The latch needle [Fig. 5 -6(a)] is so named because it can be closed using a latch which is activated without any special assistance during the knitting process. The latch needle was a more expensive and intricate needle to manufacture than the bearded needle. The old loop is cleared from the hook automatically when the needle is lifted and the hook is closed automatically after yarn feeding by lowering the needle.

钩针 (2) Spring bearded Needle

The spring bearded needle [Fig. 5 -6(b)] was the first type of needle to be produced. It is the cheapest and simplest type to manufacture as it is made from a single piece of metal. This needle has the disadvantage of requiring a pressing edge

压针板 (presser) to close the bearded hook and enclose the new loop.

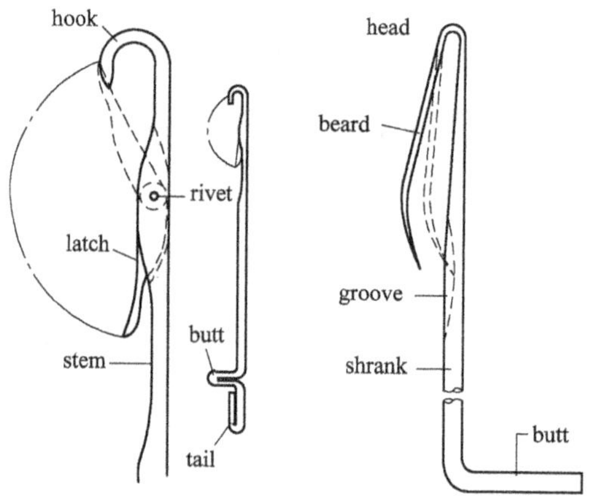

(a) Latch needle (b) Spring-bearded needle

Fig. 5 -6 knitting needles

复合针 (3) Compound Needle

The compound needle is found almost exclusively in warp knitting. This needle

针身/针芯 consists of two separately-controlled parts—the open hook and the sliding closing element (tongue, latch, piston, plunger). The two parts rise and fall as a single unit but, at the top of the rise, the hook moves faster to open the hook and at the start of the fall the hook descends faster to close the hook.

5.1.3.2 Sinker

沉降片

The sinker is the second primary knitting element. Sinkers are thin steel elements which are placed between each pair of needles. Their purpose is to control the fabric movement during needle activation. The sinker is composed of nib or nose, slot or throat, belly, and heel.

片鼻
片喉/片颚/片踵

5.2 Weft or Filling Knitting

纬编

Weft or filling knitting is a construction process in which the fabric is made by a yarn forming loops across the width or around a circle; each yarn is fed at more or less a right angle to the direction in which the fabric is formed. The term "weft" is taken from weaving techniques; in weaving, the term is used synonymously with filling to refer to the horizontal or crosswise direction of a fabric.

5.2.1 Weft Knitting Machine

纬编机

Weft knits can be made by machine or by hand. Most weft knitting is done on machines, which may be broadly classified into two types, namely circular and flat-bed machine (Fig. 5 -7, Fig. 5 -8).

A knitting machine is an apparatus for applying mechanical movement to convert yarn into knitted loop structures. The machine incorporates and co-ordinates the action of a number of mechanism and devices, each performing specific functions that contribute towards the efficiency of the knitting action. Generally speaking, knitting requires four basic steps: yarn supply, knitting elements, fabric take-down, and fabric collection.

成圈运动

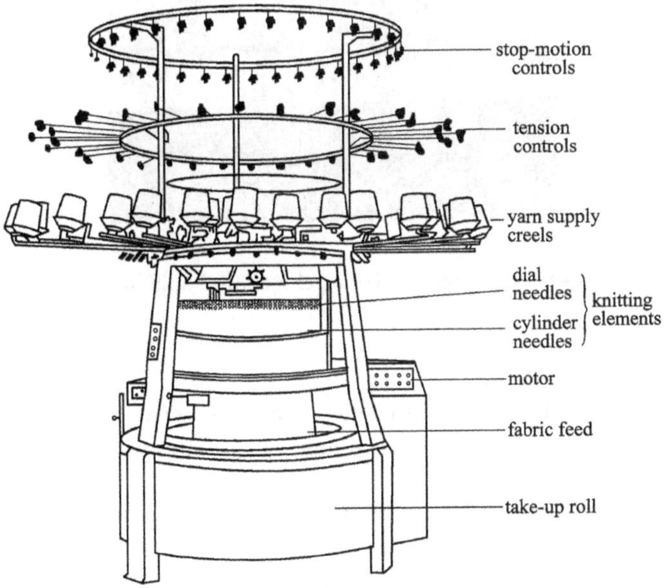

Fig. 5 -7 A circular machine

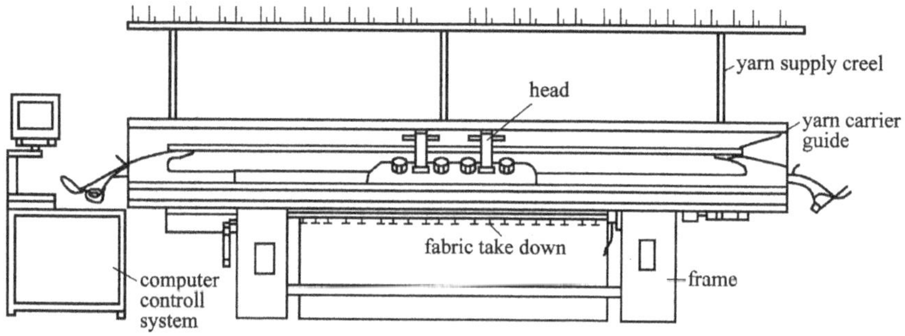

Fig. 5 – 8 A computer-controlled flat-bed knitting machine

横机

5.2.1.1 Flat-bed Machines

Flat-bed, or flat-bar, machines are characterized by the arrangement of their needles on a horizontal or flat needle bed (Fig. 5 – 9). Most flat-bed machines have two needle beds which are located opposite to each other. It is noteworthy that flat-bed knitting machines have low production rates and their fabrics are normally of large loops with low stitch densities.

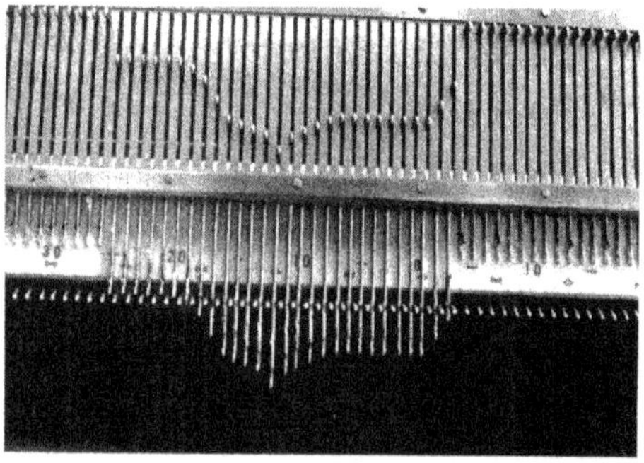

Fig. 5 – 9 Close-up of a section of a flat-bed knitting machine

圆机

5.2.1.2 Circular Machines

Most weft knitting is done on machines that have needles arranged in a circle and thus form the fabric as a tube. The tube can be slit to form a flat fabric, or it can be used as a tube.

Circular weft knitting machines may be single-bed or double-bed and their needles are arranged in a circular needle bed (Fig. 5 – 10). Single-bed machines have their needles arranged vertically along the perimeter of the circular knitting
针筒针 bed. This set of needles are called cylinder needles. Double-bed machines have an
针盘针 additional set of needles, called dial needles, mounted horizontally along the circ-

umference of a dial which in turn sits above and perpendicular to the cylinder needle bed. Higher production rates on circular weft knitting machines are achieved compared with flat-bed machines, and their fabrics normally consist of small loops with relatively high-stitch densities.

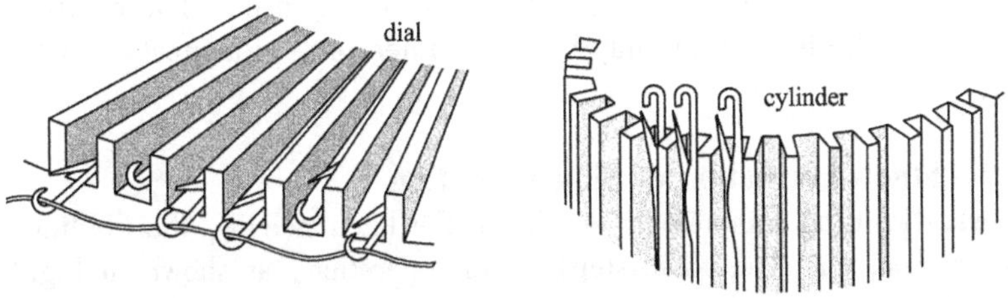

Fig. 5 – 10 Section of a dial and a cylinder with the arrangement of latch needles

5.2.2 Loops and Stitch Formation in Weft Knitting

纬编线圈及其形成

5.2.2.1 Three Types of Loops

三种线圈

In addition to knit loops, tuck and float loops also constitute very important elements of knit fabric structures. It is the judicious combination of these three elements that allows the formation of a wide range of knit fabrics.

(1) Knit Loop

线圈

The basic loop, the knit loop, is shown in Fig. 5 – 11(a). The knit loop is produced by clearing the old loop below the latch (by raising the needle) and feeding a new yarn into the hook.

(2) Tuck Loop

集圈

A tuck loop is formed if the needle is raised only partially, retaining the old loop in the hook but opening enough to accept new yarn. The tuck loop is open at the neck. The appearance of a tuck loop in the fabric is shown in Fig. 5 – 11(b).

(3) Float Loop

浮线圈

A float loop or miss stitch [Fig. 5 – 11(c)] is formed if the needle is not

(a) Basic (b) Tuck (c) Float

Fig. 5 – 11 Three types of loops

raised, preventing any yarn from being fed into it. The old loop is retained and knits on the next course but the yarn for the existing course falls behind the needle and floats on the surface.

Weft knitting and weft fabrics can be classified as single or double depending upon the number of independent needle sets required to produce them. Single knitting is weft knitting in which one set of needles is used. The needles are arranged in a needle bed which may be either linear or, as is most often the case, circular.

单面针织物线圈形成 5.2.2.2 Stitch Formation of Single Knitting

The major knitting components consist of the latch needle, the holding sinker, and the yarn carrier. The basic steps in this operation, as shown in Fig. 5 – 12, are as follows:

起始位置 (1) The needle is in the running or rest position. The last-formed loop is on the latch of the needle and the sinker is ahead holding the fabric.

(2) The needles pushed up by the cam to the uppermost or clearing position. The previous loop has slipped off the latch and lies on the stem of the needle. The sinker is still forward, holding down the formed loops. The yarn carrier feeds the yarn to the hook.

(3) After the yarn is in the hook, the needle begins to lower. The holding sinker retracts to permit the needle to drop low enough so that the yarn in the hook is pulled through the previously formed loop. The old loop closes the latch, and thus the yarn is held securely in place, ready for a new loop.

(4) The needle drops to its lowest position, forming the new loop. The sinker begins to move forward to push the old loop ahead and hold the fabric. The process is then ready to be repeated.

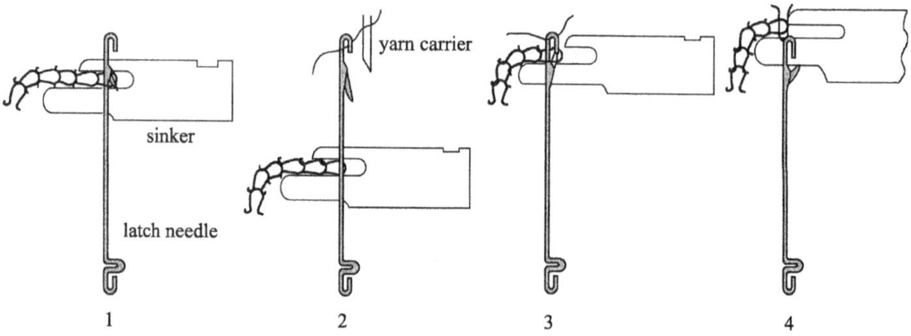

Fig. 5 – 12 Stitch formation in single knitting

双面针织物线圈形成 5.2.2.3 Stitch Formation of Double Knitting

The manufacture of double knits involves two sets of needles (i. e. , dial needles and cylinder needles) at right angles to each other. The steps in making double knits is shown in Fig. 5 – 13.

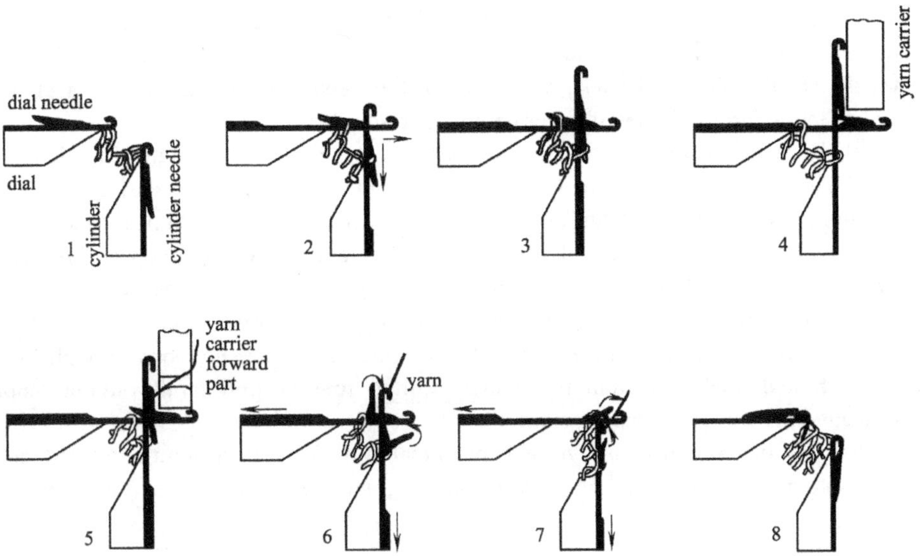

Fig. 5 – 13 Stitch formation in double knitting

5.2.3 Basic Stitches in Weft Knitting

纬编基本组织

Four primary structures—plain, rib, interlock and purl – are the base structures from which all weft knitted fabrics and garments are derived.

5.2.3.1 Plain Stitch

平针组织

The simplest of all weft knitting structures is the plain stitch or jersey stitch [Fig. 5 – 14 (a)]. It is composed entirely of face loops (or entirely of back loops). The single plain stitch, or basic structural unit, is therefore, one face loop (or one back loop).

The single plain [Fig. 5 – 14 (b)] structure exhibits the following characteristics:

平针织物

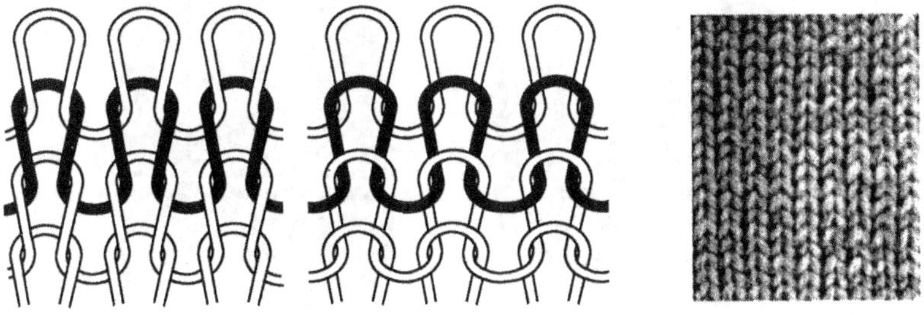

(a) The plain stitch (b) A garment knit

Fig. 5 – 14 Plain stitch

① The appearance of the face and back differ. Its technical face is smooth, with the side limbs of the needle loops having the appearance of columns of V's in the wales. On the technical back, the heads of the needle loops and the bases of the sinker loops form columns of interlocking semi-circles.

② Extensibility widthwise is approximately twice that of the length direction.

卷边 ③ The edges of the fabric tend to curl or roll. The sides will curl toward the technical back, while the top and bottom will curl toward the technical face.

④ A run (collapse of a wale) will occur if a cut or exposed loop is stressed. The direction of collapse can be either from top to bottom or vice verse.

⑤ The fabric can be unraveled, course by course, from either end. i. e. ,Plain can be unrove from the course knitted last by pulling the needle loops through from the technical back, or from the course knitted first by pulling the sinker loops through from the technical face side. 顺编织方向

Plain is the simplest and most economical weft knitted structure to produce. Plain stitch knits are used in making hosiery, sweaters, sportswear, and similar items.

5.2.3.2 Purl Stitch 双反面组织

Purl was originally spelt "pearl" and was so named because of its similar appearance to pearl droplets. The construction is also referred to as the links-and-links stitch. Purl fabrics are characterized by having both face knit loops and back knit loops in one wale.

The simplest purl, the 1 × 1 purl is illustrated in Fig. 5 – 15 and it exhibits the following characteristics:

① Same appearance, face and back (familiar to the back of single jersey).

② Highly extensible in all directions. Approximately twice as extensible as single jersey in the length direction because of the lengthwise fabric contraction, which occurs to form the coursewise ribs.

③ The fabric does not curl at the edges.

④ The fabric will run in the wale direction starting from either end.

⑤ The fabric may be unraveled course by course starting from either end.

⑥ The fabric tends to be two or three times thicker than single jersey.

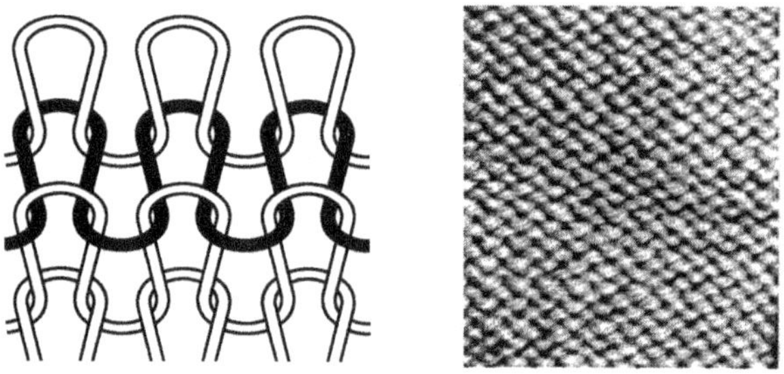

(a) The 1×1 purl stitch (b) A garment knit

Fig. 5 – 15 Purl stitch

5.2.3.3 Rib Stitch

罗纹组织

Rib structwce have alternative lengthwise rows of plain and purl stitches constructed. It consists of alternate face wales and back wales, where a face wale is composed entirely of face loops and a back wale is composed entirely of back loops.

Thus, the 1 × 1 rib structure, illustrated in Fig. 5 – 16(a), has a repeating stitch consisting of one face loop and one back loop, plus the connecting yarn. The fabric has the following characteristics:

① The appearance of the face and back are identical.

② The extensibility of the fabric widthwise is approximately twice that of single jersey. The lengthwise extensibility is essentially the same as in single jersey.

③ The fabric does not curl at the edges.

④ A run will develop in the fabric if an exposed loop is cut, as is the case for single knits, except that the direction of collapse will be from top to bottom only.

⑤ The fabric can be unraveled course by course but only from the end last knitted.

⑥ The fabric thickness is approximately twice that of single jersey.

Rib knit fabrics have a very distinct lengthwise rib effect on both sides of fabric. These are actually ridges of plain stitch wales. If two wales of plain stitches and two wales of purl stitches appear alternately on both sides of the fabric, it is called a 2 ×2 rib [Fig. 5 – 16(b)].

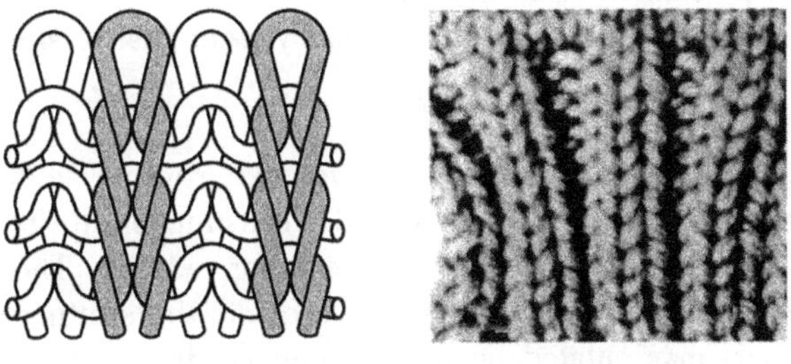

(a) The 1×1 rib stitch

(b) A garment knitted in a 2×2 rib

Fig. 5 – 16 Rib stitch

The primary disadvantage of rib knits is high cost due to the increase in fabric weight and use of yarn and the relatively low output of the machine. Rib knit fabrics are usually found at the lower edge of sweaters, on sleeves, and at necklines.

5.2.3.4 Interlock Stitch

双罗纹组织

The interlock stitch is a variation of the rib stitch. The simplest fabric produced on an interlock gated machine is called interlock shown in Fig. 5 – 17(a). Close examination of this structure reveals that it is composed of two 1 × 1 rib fabrics locked together. The characteristics of an interlock fabric [Fig. 5 – 17(b)] are summarized below:

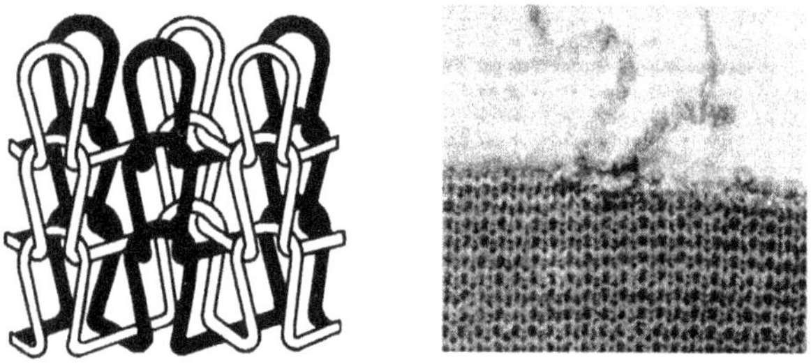

(a) The interlock stitch (b) A garment knit

Fig. 5 – 17 Interlock stitch

① The appearance of the face and back is the same.

② Extensibility widthwise and lengthwise is approximately the same as single jersey.

③ The fabric does not curl at the edges.

④ A run will develop in the fabric in the direction of which will be from the end last knitted. An interlock fabric will run less freely than single jersey or rib structures.

⑤ The fabric can be unraveled from the end last knitted. Two yarns must be removed to unravel a complete knitted course.

⑥ The thickness of the fabric is approximately twice that of single jersey.

5.3 Warp Knitting

经编

In warp knitting the loops are formed in a vertical or warpwise direction. All the yarns required for the width of the fabric under construction are placed parallel to each other on a beam that resembles a warp beam for weaving. This beam is set into the warp knitting machine, and all yarns feed into the knitting area simultaneously. Each yarn is manipulated by one specific needle, and all the yarns form loops at the same time.

经轴

5.3.1 Warp Knitting Machines

经编机

特里科经编机/拉舍尔经编机

The two major classes of warp knitting machines are Tricots and Raschels. In the past, tricot machines mainly employed bearded needles with a presser bar and Raschels used latch needles together with a latch wire or blade. Now, despite wide spread use of the compound needle, there are still distinctive differences between the two types.

5.3.1.1 Tricot Machine

特里科经编机

A Tricot machine is a warp knitting machine which uses a single set of springbeard or compound needles. The fabric is removed from the needles at approximately 90 degrees. Fig. 5 – 18 (a) shows a cross section of the knitting elements found on a Tricot machine.

5.3.1.2 Raschel Machine

拉舍尔经编机

A Raschel machine is a warp knitting machine which uses a single set of vertically mounted latch needles. The fabric is removed from the needles at an angle of 120 - 160 degrees. Fig. 5 - 18(b) shows a cross section of the knitting elements found on a Raschel machine.

Unlike the Tricot knit machines, which work best with comparatively fine yarns, raschel machines use any conceivable type of yarn. It is especially suitable to heavy yarns and coarse-gauge designs.

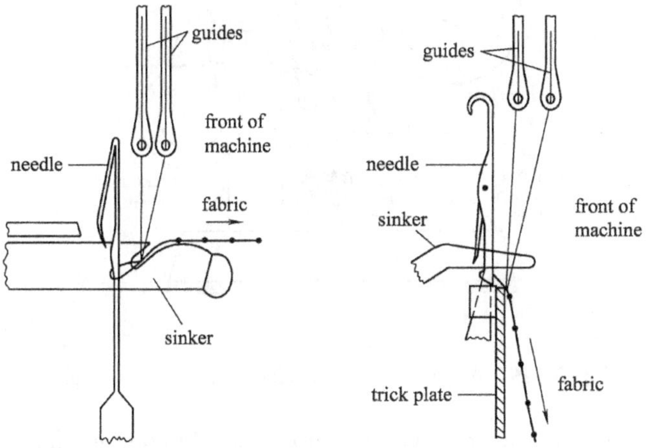

(a) A bearded needle Tricot machine (b) A latch needle Raschel machine

Fig. 5 - 18 Elements in warp knitting

5.3.2 Laps and Stitch Formation in Warp Knitting

经编线圈及其形成

Tricot machine in Fig. 5 - 19 is illustrated to detail the knitting elements of warp knitting machine and the lap formation in warp knitting.

5.3.2.1 Warp Knitting Elements

经编针织机件

Apart from needles and sinkers, three terms—warp beams, guide bars and guides should be introduced in warp knitting.

(1) Warp Beams

经轴

Yarn is supplied to the needles in the form of warp sheets. Each individual warp sheet is usually supplied from its own beam, which may consist of several section beams. The number of beams used on a machine is normally equal to the number of guide bars.

(2) Guide Bars

梳栉

The guide bars extend across the complete width of the machine and their function is to wrap yarn around the needles (i. e. , feed). Each guide in the guide bar

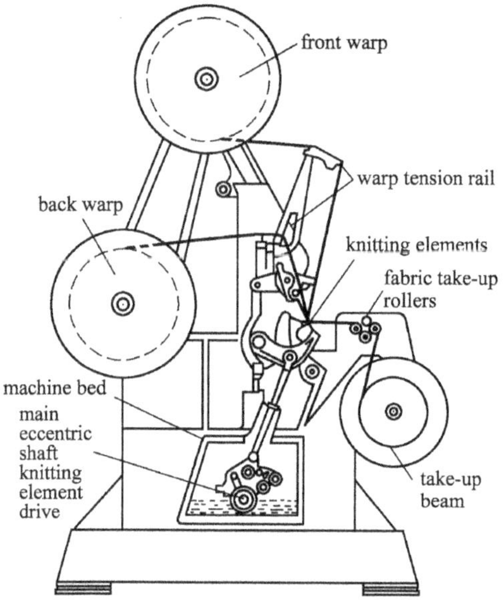

Fig. 5 – 19 A typical tricot machine. The front is at right

is usually provided with a single end of yarn. Warp knitting machines are usually equipped with two or more guide bars. Each guide bar has its own warp beam.

导纱针 (3) Guides

Warp guides are thin metal plates drilled with a hole in their lower ends through which warp ends may be threaded if required. They are held together at their upper 机号 ends as the single units in a metal lead and are spaced to the same gauge as the needles.

特里科经编机编织循环 5.3.2.2 Knitting Cycle of the Tricot Machine

Tricot fabrics, like weft knits, can be made with either latch or beard needles. Beard needles are the most common for warp knitting. Therefore the steps for making tricot are diagramed showing beard needles. The steps, as shown in Fig. 5 – 20, are as follows:

① The following are identified: warp end or yarn, guide bar that controls the yarn, presser bar, sinker; beard needles have a spring type of hook without a latch. At this point the needles are in their highest position, the previous loops are on the stems of the needles, and the sinker is holding down the formed fabric.

② The guide bar carrying the yarn that formed the last loop on the left needle moves to the right to place the yarn in the needle to the right. A yarn from the far left is being placed in the left needle.

③ The yarn is against the needle stems, and the needles start to descend. As the needles move down, the yarn will be trapped in the hooks. The presser bar is

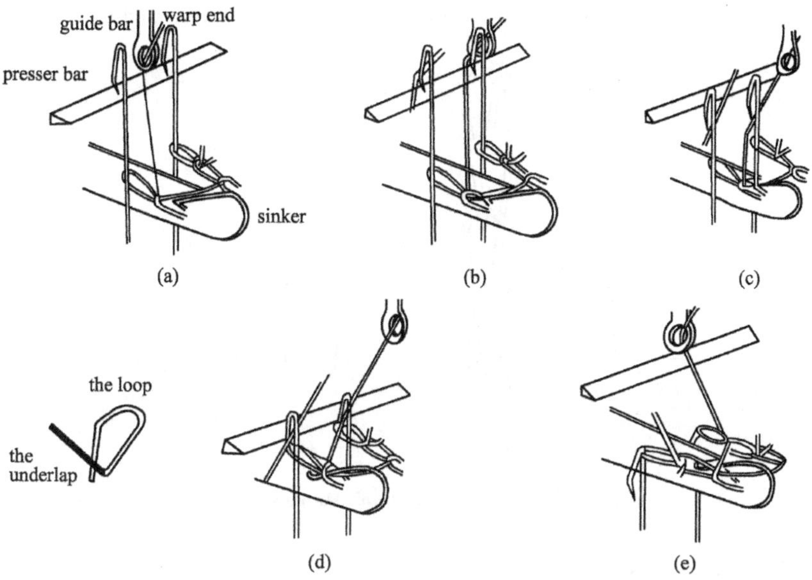

Fig. 5 – 20 Stitch formation in tricot knitting

moving toward the needles, and the sinker moves away.

④ The yarn is in the hooks. The presser bar moves in and forces the tips of the needles against the stems to hold the yarn and to permit the needles to pull the yarn through the previous loops.

⑤ The needles move to the lowest position, the new loops are formed and the old loops knock off. The sinker moves back in to hold the fabric. The needles will return to the upper position, and the steps will then be repeated.

5.3.3 Basic Stitches in Warp Knitting 经编的基本组织

When using either open or closed laps there are three possible arrangements of lapping at successive courses—pillar, tricot and atlas, which may be used alone or in combination.

5.3.3.1 Pillar Stitch 编链组织

In the pillar or chain stitch, the same guide always overlaps the same needle. (编链组织) This lapping movement will produce chains of loops in unconnected wales, which (线圈链) must be connected together by the underlaps of a second guider bar.

Open-lap pillar stitches[Fig. 5 – 21(a)] are commonly used in warp knitting. (开口编链) They can be unrove from the end knitted last. Closed-lap pillar stitches[Fig. 5 – 21(b)] are employed on crochet machines because the lapping movement is simple to achieve and is necessary when using self-closing carbine needles, which must be fed with yarn from the same side. (闭口编链; 自闭钩针)

Generally, pillar stitches are made by front guide bars, either to produce vertical (前梳栉/纵条纹效应)

stripe effects or to hold the inlays of other guide bars into the structure.

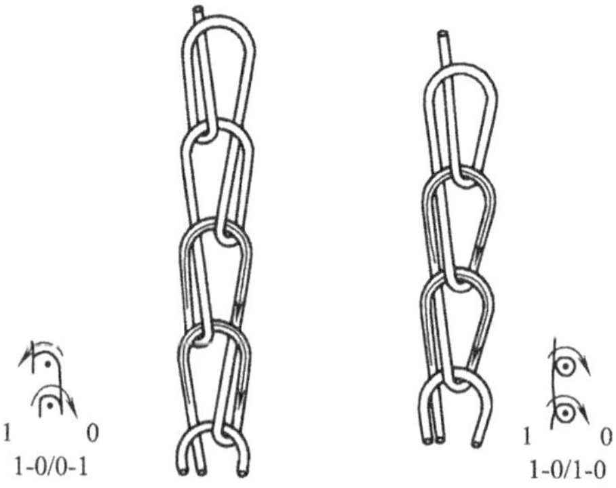

(a) Open (b) Closed

Fig. 5 – 21 Lap pillar stitches

经平组织

5.3.3.2 Tricot Stitch

经平垫纱

Tricot lapping or 1 × 1 is formed by the guide lapping alternately two adjacent needles. The definition"1 × 1" implies an underlap of one needle space and one overlapped needle. As with pillar lapping, the tricot lapping movement can be formed open [Fig. 5 – 22(a)] or closed [Fig. 5 – 22(b)]. The closed tricot fabric shown in Fig. 5 – 23 is more popular.

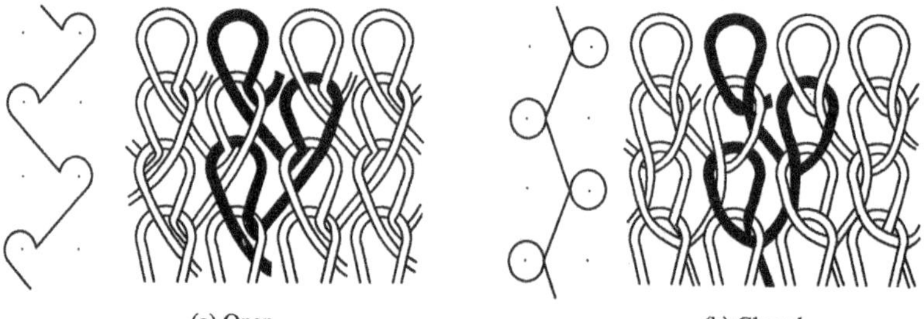

(a) Open (b) Closed

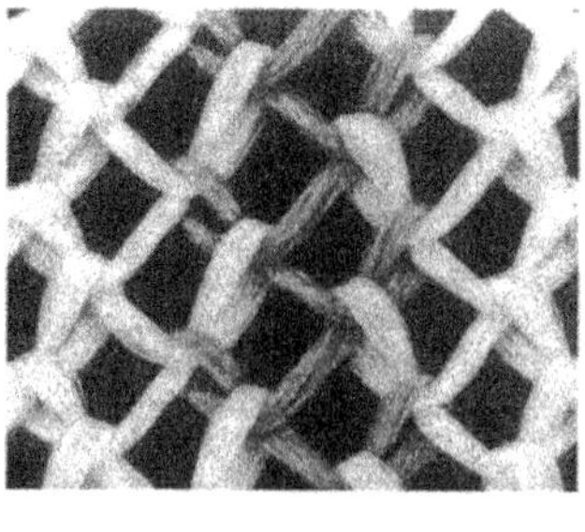

Fig. 5 – 23 Tricot stitch fabric

5.3.3.3 Atlas Stitch

经缎组织

Atlas stitch (Fig. 5 – 24) is a movement where the guide bar laps progressively in the same direction for a minimum of two consecutive courses, normally followed by an identical lapping movement in the opposite direction. Usually, the progressive lapping is in the form of open laps and the change of direction course is in the form of a closed lap.

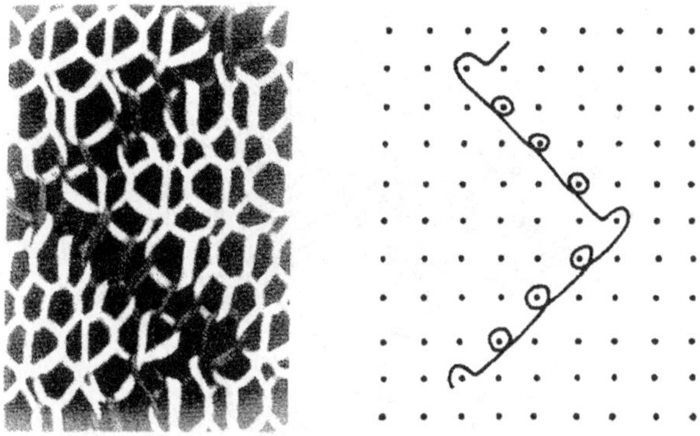

Fig. 5 – 24 Atlas stitch

5.3.3.4 In-lay Yarns and Float Stitches

衬纱组织

Weft-insert, weft-knit fabrics [Fig. 5 – 25(a)] are produced on flat-bed machines. Alternative to weft-insert, weft-knit fabrics, straight, inlay yarns can also be introduced into warp-knit structures using Raschel machines. A warp-insert, warp-knit fabric has a typical warp-knit structure but between these wales are laid unlooped yarn with minimum crimp in them [Fig. 5 – 25(b)].

(a) A typical weft-insert, weft-knit fabric

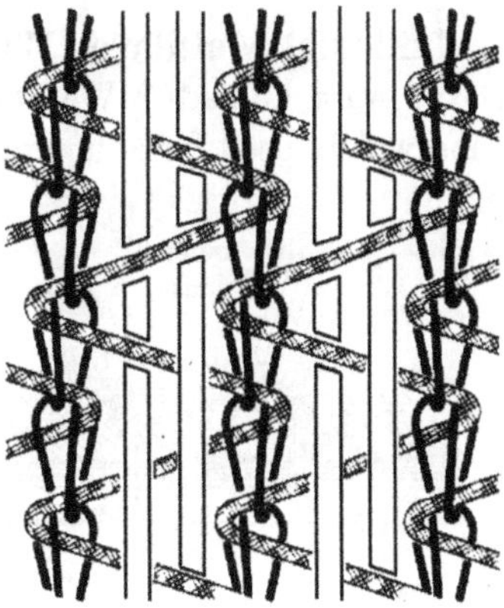

(b) A warp-insert, warp-knit fabric

Fig. 5 – 25 In-lay Yarns and float stitchs

Weft-insert, warp-knit fabrics are in principle produced in a way similar to that employed for weft-insert, weft-knit fabrics, except in this case a warp knitting machine (Fig. 5 - 26) is used instead of a weft.

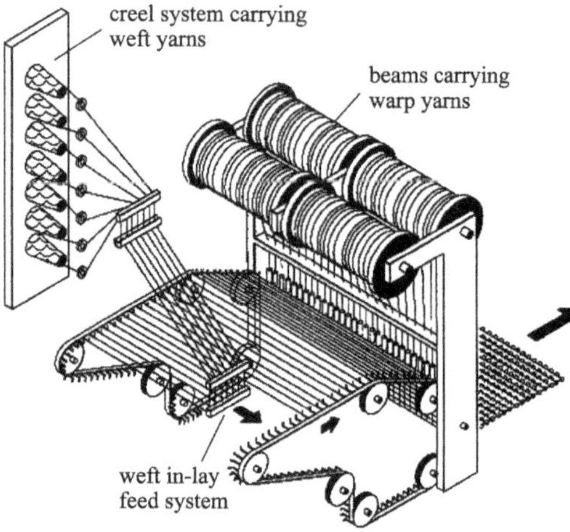

Fig. 5 - 26 Schematic diagram showing the general principle of weft-insert warp knitting on a raschel machine

三维针织预制件

5.4 Three-Dimentional Knitted Preforms

Woven and braided fabrics have been extensively used in composite materials for many years. The use of knitted fabrics as reinforcing materials emerged only one and half decades ago. And until recently knitted fabrics are considered for applications in polymer composites.

多轴向多层(非屈曲)经编织物

5.4.1 Multiaxial Multilayer Warp-knit(non-crimp)Fabrics

Multi-axial warp knitted(MWK) fabrics(Fig. 5 - 27) are a family of technical textiles, and widely used for structural composites.

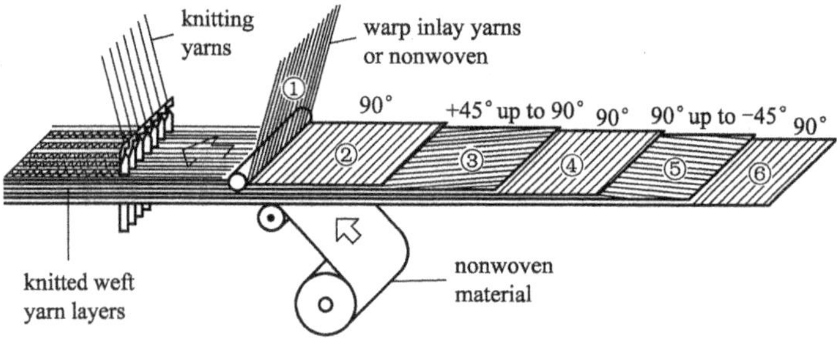

Fig. 5 - 27 Principle of the LIBA multi-axial magazine weft insertion warp knitting machine up to 6 layers and one fleece layer are possible.

5.4.2 Sandwich Structure Preforms

夹层结构预制件

3D knitted sandwich fabrics [Fig. 5 – 28(a)] were developed as reinforcement for polymer composites.

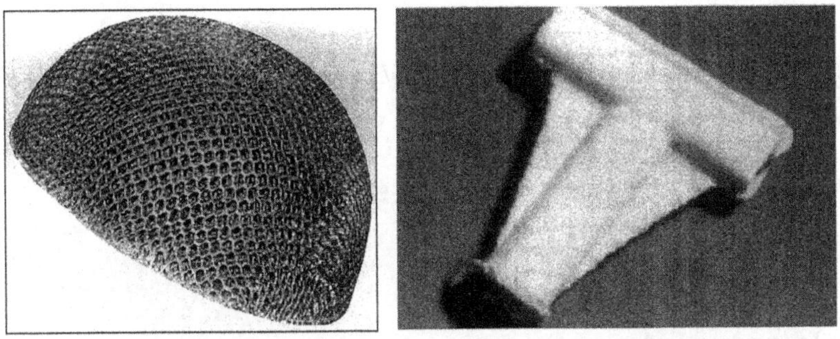

(a) 3D knitted sandwich preform for a bicycle helmet

(b) Fully fashioned glass preform for stiffened T-joints.

Fig. 5 – 28 Examples of knitted composites

5.4.3 Fully Fashioned Preforms

全成形预制件

The fully fashioned knitting technology has been used to produce near-net-shape reinforcement for engineering composites, but as yet only at demonstration levels. Fig. 5 – 28(b) shows one demonstration component of T-shape connectors.

Review Questions

思考题

1. In what respects does the knitted construction differ from the woven one? Consider the yarns, construction and operation steps.
2. Give out the methods to express knitting structures and principles.
3. What is the chief advantage gained by the knitted construction?
4. Give out the main parts of knitting machine?
5. Tell the difference among latch, spring beard and compound needle?
6. Distinguish between a miss and a tuck stitch.
7. Tell the main types of weft knitting machines?
8. What are the three principal stitches in weft knitting? What are the characteristics of the basic types of weft knits?
9. (1) Describe warp knitting.
   (2) What advantages has warp knitting over weft knitting?
10. What are the characteristics of the basic types of warp knits?

Suggested Activities

实训题

Identify five different kinds of knitting construction found in your household that are used for apparel or home furnishings. Describe their construction and appearance that led to each identification.

Chapter 6 Nonweaving and Nonwoven Fabrics

非织造与非织造织物

引言 6.1 Introduction

Nonwoven fabrics usually refer to those fabrics that have not been produced by conventional methods such as weaving, knitting, and so on. A fiber web is bonded by various means to form an integral fabric structure in nonwoven. However, there is no internationally agreed definition of nonwovens, in spite of the fact that the International Standards Organization published a definition in 1988 (ISO 9092: 1988). Many authorities think the definition of ISO 9092 is very narrow, because felts, needled fabrics, and stitch-bonded materials may not strictly be described by the definition. 毛/针刺非织造布/缝编

"A nonwoven is a textile structure produced by the bonding or interlocking of fibers, or both, accomplished by mechanical, chemical, thermal or solvent means and combinations thereof. The term does not include paper or fabrics that are woven, knitted or tufted." This is the definition of nonwoven fabrics by the American Society for Testing Materials (ASTM D 1117—80). This definition includes many important fabrics which most people regard as nonwovens, but which are excluded by ISO 9092.

One of the major advantages of nonwoven fabrics is their speed of manufacture, which is usually much faster than all other forms of fabric production, for example, spunbonding can be 2,000 times faster than weaving. This naturally means that the manufacture of nonwoven fabrics is very economical, and the cost is very low. However, they are also very versatile materials. A variety of different physical forms with a wide range of properties can be produced by different fibers with different technique and binders. 纺黏法

Almost all nonwoven fabrics are manufactured by two main steps. The first one is web formation, in which fiber is processed to form a fiber web. There are a number of different ways of web formation, and each way will produce its own particular characteristic in the final fabric. The second one is bonding of the (fiber) web, in which web fibers are bonded together. Equally there are a number of different ways of bonding of the web, and they will have an even bigger effect on the finished fabric. Almost every web formation way can be combined with every bonding way, so that the possible manufacturing lines of nonwoven fabrics, and there- 纤维成网/纤维网 纤维网结合 生产线

fore the final nonwoven fabric products are enormous. Nonwoven fabric products are often named according to the two main manufacture steps. Dry-laid, wet-laid, melt-blown, needlepunched, spunbonded, stitch-bonded, and spunlaced are typical nonwoven fabrics. 针刺法非织造布/纺黏法非织造布

6.2 Web Formation 纤维成网

6.2.1 Dry-laying 干法成网

There are three methods in dry-laying: parallel-laid, cross-laid and air-laid. They account for slightly less than half the total nonwoven production. The three dry-laid processes are somewhat similar to traditional textile processing. 平行铺网/交叉铺网/气流铺网

6.2.1.1 Parallel-Laid 平行铺网

In parallel-laid, each layer of fiber is produced by carding. The principles of carding and the types of carding machines are quite similar to yarn manufacture. The mass and strength of each layer of fiber is normally too low. Several layers of fiber are laid parallel, so named "parallel-laid". Additionally, the uniformity can be increased by laying several carded fiber layers over each other to form the web. Fig. 6 – 1 shows three cards raised slightly above the floor to allow a long conveyor to pass underneath. The layers of fiber from each card fall onto the conveyor forming a web with three times the mass of each layer of fiber. 纤维层

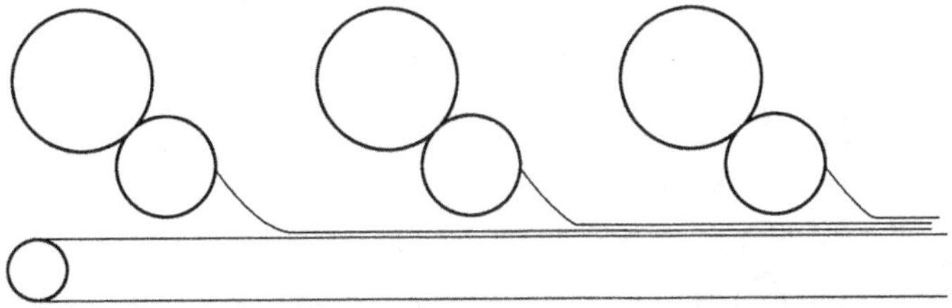

Fig. 6 – 1 Diagram of parallel-laid with three layers of fiber

However, there are two problems in the parallel-laid. The first one is the weakness in cross direction. In nonwoven, the direction along the web is called the "machine direction" and the direction at right angles is called the "cross direction". This problem can be altered by the special designed cards, which are given a randomising doffer, together with scrambling rollers. The other problem is the width of final nonwoven fabric. The width of the final fabric cannot be wider than the carded layer of fiber, while current trends demand wider and wider fabrics. 机器方向/宽度方向 随机落纱器/不规则罗拉

6.2.1.2 Cross-Laid 交叉铺网

In cross-laid, each layer of fiber is also produced by carding. However, the cards are placed at right angles to the main conveyor, and the carded layers of fiber are traversed backwards and forwards across the main conveyor. The problems of the weakness in cross direction and the limited width do not exist in cross-laid.

There are another two major problems with cross layers; one is that they tend to lay the web heavier at the edges than in the middle. This problem can be corrected by running the traversing mechanism rather slower in the center and more rapidly
匹配 at the edges. The other problem is trying to match the input speed of the cross layer with the card web speed. For various reasons the input speed of the cross layer is limited and the speed of the card web has to be reduced to match it.

气流铺网 6.2.1.3 Air-Laid

The air-laid method produces the final web in one stage without first making the layer of fiber. It is also capable of running at high production speeds. But it is similar to the parallel-laid method in that the width of the final web is limited.

The diagram in Fig. 6 – 2 shows the principle of an air-lay machine. Opened fiber from the opening/blending section is fed into the back of hopper A, which delivers a uniform sheet of fibers to the feed rollers. The fiber is then taken by the
刺辊 toothed roller B, which is revolving at high speed. There may or may not be worker and stripper rollers set to the roller B to improve the opening power. A strong air
移走 stream C dislodges the fibers from the surface of roller B and carries them onto the conveyor F, on which the web is formed. The stripping rail E prevents fiber from recirculating round the cylinder B. The air flow at D helps the fiber to stabilize in the formation zone.

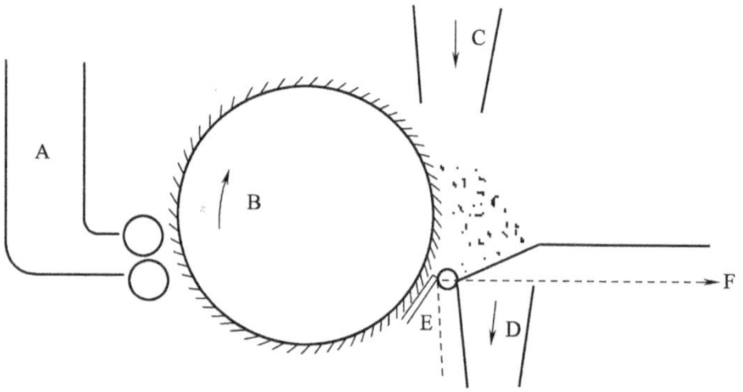

Fig. 6 – 2 Diagram of an air-lay machine

湿法成网 6.2.2 Wet-laying

The main advantage of the wet-laying is its high production speeds. The wet-lay-
造纸 ing process shown in Fig. 6 – 3 is a development from papermaking. Textile fibers are cut very short by textile standards (6 ~ 20mm), but at the same time these are
木浆 very long in comparison with wood pulp, the usual raw material for paper. The fibers are dispersed into water; the rate of dilution has to be great enough to prevent
聚集 the fibers aggregating. The required dilution rate turns out to be roughly ten times that required for paper, which means that only specialized forms of paper machines can be used. In fact most frequently a blend of textile fibers together with wood

pulp is used as the raw material, not only reducing the necessary dilution rate but also leading to a big reduction in the cost of the raw material. It is now possible to appreciate one of the problems of defining "nonwoven". It has been agreed that a material containing 50% textile fiber and 50% wood pulp is a nonwoven, but any further increase in the wood pulp content results in a fiber-reinforced paper. A great many products use exactly 50% wood pulp. Wet-laid nonwovens represent about 10% of the total market, but this percentage is tending to decline. They are used widely in disposable products, for example in hospitals as drapes, gowns, sometimes as sheets, and in disposable nappies.

手术巾/手术衣
尿布

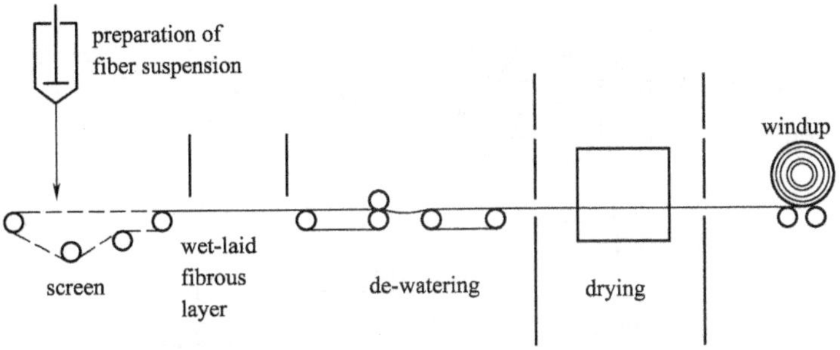

Fig. 6 – 3 Diagram of wet-laying

6.2.3 Spun laying

纺丝成网

Spun laying includes extrusion of the filaments from the polymer raw material, drawing the filaments and laying them into a web. As laying and bonding are normally continuous, this process represents the shortest possible textile route from polymer to fabric in one stage. In addition to this the spun laying process has been made more versatile. When first introduced only large, very expensive machines with large production capabilities were available, but much smaller and relatively inexpensive machines have been developed, permitting the smaller nonwoven producers to use the spun laying route. Further developments have made it possible to produce microfibers on spun laying machines giving the advantages of better filament distribution, smaller pores between the fibers for better filtration, softer feel and also the possibility of making lighter-weight fabrics. For these reasons spun-laid production is increasing more rapidly than any other nonwoven process.

Spun laying starts with extrusion. Polyester and polypropylene are by far the most common, but polyamide and polyethylene can also be used. The polymer chips are fed continuously to a screw extruder which delivers the liquid polymer to a metering pump and then to a bank of spinnerettes, or alternatively to a rectangular spinnerette plate, each containing a very large number of holes. The liquid polymer pumped through each hole is cooled rapidly to the solid state but some stretching or drawing in the liquid state will inevitably take place. Up to this stage the technology is similar to the fiber or filament extrusion in synthetic fiber production,

螺杆挤出机
计量泵

except that the speeds are higher, but here the technologies tend to divide. In spun laying the most common form of drawing the filaments to obtain the correct modulus is air drawing (气流拉伸), in which a high velocity air stream is blown past the filaments moving down a long tube, the conditions of air velocity and tube length being chosen so that sufficient tension is developed in the filaments to cause drawing to take place. In some cases air drawing is not adequate and roller drawing has to be used as in normal textile extrusion, but roller drawing is more complex and tends to slow the process, so that air drawing is preferred.

A simplified diagram of the spun laying is shown in Fig. 6 –4. The spunbonded nonwoven is produced by spun laying with thermal bonding.

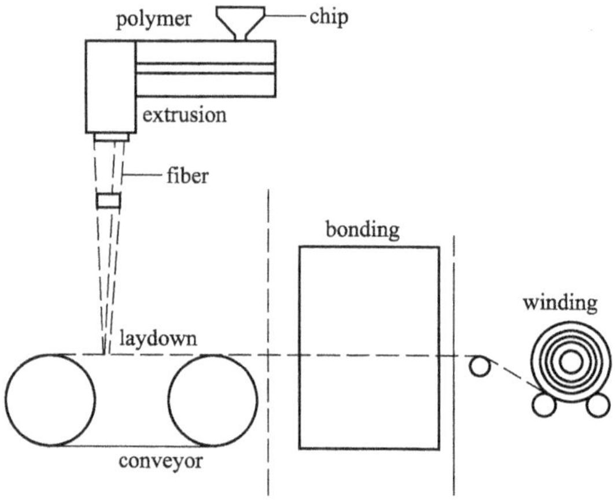

Fig. 6 –4 Diagram of the spun laying

6.2.4 Melt-blown (熔喷成网)

The process of melt-blown is another method of producing very fine fibers at high production rates. Fig. 6 –5 shows that the diagram of the melt-blown.

As the polymer leaves the extrusion holes it is hit by a high speed stream of hot

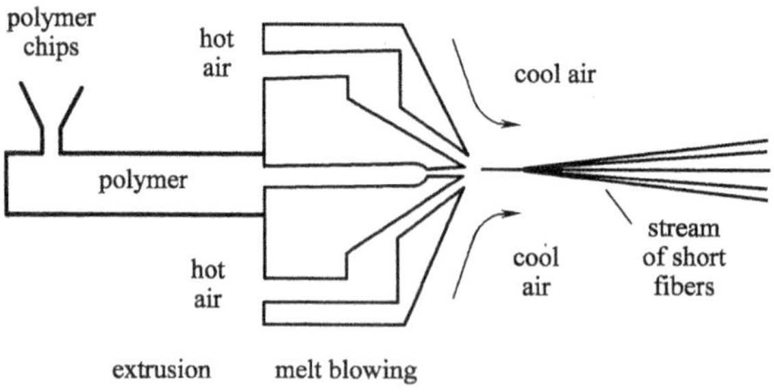

Fig. 6 –5 Diagram of the melt-blown

air at or even above its melting point, which breaks up the flow and stretches the many filaments until they are very fine. At a later stage, cold air mixes with the hot and the polymer solidifies. Depending on the air velocity after this point a certain amount of aerodynamic drawing may take place but this is by no means as satisfactory as in spun laying and the fibers do not develop much tenacity. At some point the filaments break into staple fibers, but it seems likely that this happens while the polymer is still liquid because if it happened later this would imply that a high tension had been applied to the solid fiber, which would have caused drawing before breaking. The fine staple fibers produced in this way are collected into a web on a conveyor as in air laying and spun laying. The big difference is that in melt blowing the fibers are extremely fine so that there are many more fiber-to-fiber contacts and the web has greater integrity.

固化剂
空气动力学的

For many end-uses no form of bonding is used and the material is simply a web of loose fibers. Such uses include ultrafine filters for air conditioning and personal face masks, oil-spill absorbents and personal hygiene products. In other cases the melt-blown web may be laminated to another nonwoven, especially a spun-laid one or the melt-blown web itself may be bonded but the method must be chosen carefully to avoid spoiling the openness of the very fine fibers. In the bonded or laminated form the fabric can be used for breathable protective clothing in hospitals, agriculture and industry, as battery separators, industrial wipes and clothing interlinings with good insulation properties.

个人卫生产品
电池隔离物

6.3 Bonding of the Web

纤维网结合

There are three general ways of bonding webs together: chemical bonding, thermal bonding, and mechanical bonding.

黏合剂法/热黏合法

In mechanical bonding, some forms of mechanical treatment such as needling, water jet, or stitching are used. There are three kinds of methods for mechanical bonding: needlepunching, hydroentanglement, and stitch bonding. The method chosen depends on the actual properties required in the end-product, the price structure and the equipment available.

机械结合法
水射流/缝编
针刺法/水刺法/缝编法

6.3.1 Chemical Bonding

黏合剂法

Chemical bonding is usually carried out by treating the web with a binder applied by aqueous impregnation, foam processing, coating, printing, powder adhesive scattering or by a solvent-based or water-based spraying method. The binders are selected depending on the end-use, durability requirements, cost or downstream processing considerations. All the common chemical types are used, including acrylic, acrylic copolymer, SBR, vinyl acetate ethylene copolymer etc., and application of the binder is followed by drying, and curing, if the resin is a cross-linking type.

黏合剂
浸渍法/泡沫法/涂层法/喷洒法

Chemical bonding involves treating either the complete web or alternatively isolated portions of the web with a bonding agent with the intention of sticking the fibers together. Although many different bonding agents could be used, the modern industry uses only synthetic latex, of which acrylic latex represents at least half and

合成乳液/丙烯酸乳液

丁苯乳液/醋酸乙烯乳液 styrene – butadiene latex and vinyl acetate latex roughly a quarter each. When the bonding agent is applied it is essential that it wets the fibers, otherwise poor adhesion will be achieved. The next stage is to dry the latex by evaporating the aqueous component and leaving the polymer particles together with any additives on and between the fibers. The final stage is curing and in this phase the web is brought up to a higher temperature than for the drying. The purpose of curing is to develop crosslinks both inside and between the polymer particles and so to develop good cohesive strength in the binder film.

饱和浸渍黏合 Saturation bonding wets the whole web with bonding agents, so that all fibers are covered in a film of binder. To saturate the web, it is carried under the surface of the bonding agent. In most cases the web is very open and weak and care is needed to avoid distortion. The action of the liquid greatly reduces the thickness of the web and the thickness is further reduced by the squeeze rollers which follow. Hence saturation bonded fabrics are generally compact and relatively thin.

The drying of saturation bonding nonwovens is often done in a dryer designed basically for woven fabric, modified by having a permeable conveyor to support and transport the nonwoven through the machine. Hot air is blown against the top and bottom surfaces to cause drying. Because the air only penetrates the immediate surfaces of the nonwoven, drying is confined to these areas and the central layers of the nonwoven remain wet. The result is that the liquid wicks from the wet areas to the dry ones, unfortunately carrying the suspended binder particles with the water. 黏合剂迁移 This is the cause of the problem known as binder migration; under adverse conditions virtually all the binder can be found in the surface layers and the central layers contain no binder at all, leaving them very weak. Such a fabric can easily split 分层 into two layers, a problem known as delamination. Fortunately a number of ways have been found to control binder migration, several of them being rather complicated.

It can be inferred above that one of the problems of saturation bonding is that too much water is used. This not only increases the cost of drying but also increases the risk of binder migration. Application of chemicals as a foam was developed not only for nonwovens but also for the dyeing and finishing industry as a means of using less water during application. The binder solution and a measured volume of air are passed continuously through a driven turbine which beats the two components into a consistent foam. The foam is then delivered to the horizontal nip of the 泡沫黏合 impregnating roller. Foam bonding should only be thought of as an alternative method of saturation bonding. The properties and uses of the fabrics are identical.

印花黏合 Print bonding involves applying the same types of binder to the web but the application is to limited areas and in a set pattern. The binder does not penetrate well into the dry web so the web is first saturated with water and then printed with either a printing roller or a rotary screen printer. Print-bonded fabrics are much softer in feel and also much more flexible owing to the strong effect of the free fibers in the unbonded areas. They are also significantly weaker than saturation-bonded fabrics owing to the fibers slipping in unbonded areas, but knowing the fiber length and the fiber orientation distribution it is possible to design a print pattern which will minimize the strength loss. Print-bonded fabrics tend to be used in applications where the textile-like handle is an advantage. Examples are disposable/protective clothing, coverstock and wiping cloths, particularly domestic ones. 面层材料/揩布

Similar latex binders may also be applied by spraying, using spray guns similar to those used in painting, which may be either operated by compressed air or be airless. The final product of spraying bonding is a thick, open and lofty fabric used widely as the filling in quilted fabrics, for duvets, for some upholstery and also for some types of filter media. 喷枪 喷洒黏合 垫子

6.3.2 Thermal Bonding 热黏合法

Thermal bonding makes use of the thermoplastic nature of the web fiber itself, or an external hot-melt adhesive is sometimes introduced. Quite often a lower melting point bicomponent fiber is made and co-spun and this bicomponent fiber acts as the binder. Bonding is achieved using a calendar process, by controlled hot air, infra-red heaters or by high frequency or ultrasonic welding. Thermal bonding is increasingly used at the expense of chemical bonding for a number of reasons. Thermal bonding can be run at high speed, whereas the speed of chemical bonding is limited by the drying and curing stage. Thermal bonding takes up little space compared with drying and curing ovens. Also thermal bonding requires less heat compared with the heat required to evaporate water from the binder, so it is more energy efficient. 双组分纤维

Thermal bonding can use three types of fibrous raw material, each of which may be suitable in some applications but not in others.

① The fibers may be all of the same type, with the same melting point. This is satisfactory if the heat is applied at localised spots, but if overall bonding is used it is possible that all the fibers will melt into a plastic sheet with little or no value.

② A blend of fusible fiber with either a fiber with a higher melting point or a non-thermoplastic fiber can be used. This is satisfactory in most conditions except where the fusible fiber melts completely, losing its fibrous nature and causing the web to collapse in thickness. 易熔纤维

③ The fusible fiber may be a bicomponent fiber, that is, a fiber extruded with a core of high melting point polymer surrounded by a sheath of lower melting point polymer. This is an ideal material to process because the core of the fiber does not melt but supports the sheath in its fibrous state. 外壳,鞘

Thermal bonding is used with all the methods of web production except the wet-laid method, but it is worth pointing out that the spun-laid process and point bonding of thermal bonding complement each other so well that they are often thought of as the standard process. 点黏合

Thermal bonding can be processed without pressure. In this case, the web is processed through a hot air oven with just sufficient air movement to cause the fusible portion to melt. This method is used to produce high loft fabrics. The products are similar to spray-bonded materials except that in this case the bonding is uniform all the way through and there is virtually no limit to the thickness of fabric made. The uses of the thermal-bonded fabric are basically the same as those of a spray-bonded fabric but they would be used in situations where a higher specification is

燃料箱/一级方程式赛车 required. One interesting example is in the fuel tanks of Formula 1 cars to limit the rate of fuel loss in the event of a crash.

In other cases, thermal bonding can be processed with some pressure. This method is basically the same as the previous one, except that as the web leaves the thermobonding oven it is calendered by two heavy rollers to bring it to the required thickness. The products could be used as insulation, as rather expensive packaging or for filtration.

Thermal bonding can also be processed with high pressure. Here, the web is taken between two large heated rollers (calender rollers) that both melt the fusible fiber and compress the web at the same time. Provided the web is not too heavy in mass per unit area the heating is very rapid and the process can be carried out at high speed (300 m/min). The products tend to be dense and heavily bonded, although of course the amount of bonding can be adjusted by varying the percentage of fusible fiber in the blend. Typical properties are high strength, very high modulus and stiffness but good recovery from bending. The main uses are in some geotextiles, stiffeners in some clothing and in shoes, some filtration media and in roofing membranes. 屋顶材料

Although it is very strong, the fabric produced with bonding all over the web (area bonding) is too stiff and non-textile-like for many uses. It is far more common to use point bonding, in which one of the calender rollers is engraved with a pattern that limits the degree of contact between the rollers to roughly 5% of the total area. The bonding is confined to those points where the rollers touch and leaves roughly 95% of the web unbonded. The area, shape and location of the bonding points are of great importance. Fabrics made in this way are flexible and relatively soft owing to the unbonded areas. At the same time they maintain reasonable strength, especially in the case of the spun-laid fabrics. These fabrics have many uses, for example, as a substrate for tufted carpets (簇绒地毯底布), in geotextiles, as a filtration medium, in protective/disposable clothing, as a substrate for coating (涂层底布), in furniture and home furnishings and as coverstock.

6.3.3 Needlepunching 针刺法

Needlepunching is extensively used as a means of mechanically bonding webs together. Barbed needles (带钩刺针) are inserted into the web which force some of the fibers downward causing them to be entangled with fibers in lower layers of the web. Further entanglement (缠结) occurs on the upward movement of the needles. A wide variety of properties can be obtained using specially designed needles and by joining together webs of different characteristics. Non-woven fabrics with the characteristics of both weaving and knitting and the appearance and properties of fleece and velours (绒头织物/天鹅绒) can be produced by needlepunching.

The basic concept of needlepunching is apparently simple. The web is led between two stationary plates, the bed and stripper plates (托网板和剥取板), as shown in Fig. 6 - 6. While between the plates the web is penetrated by a large number of needles (刺针). The needles are usually made triangular and have barbs (倒钩) cut into the three edges (see Fig. 6 - 7).

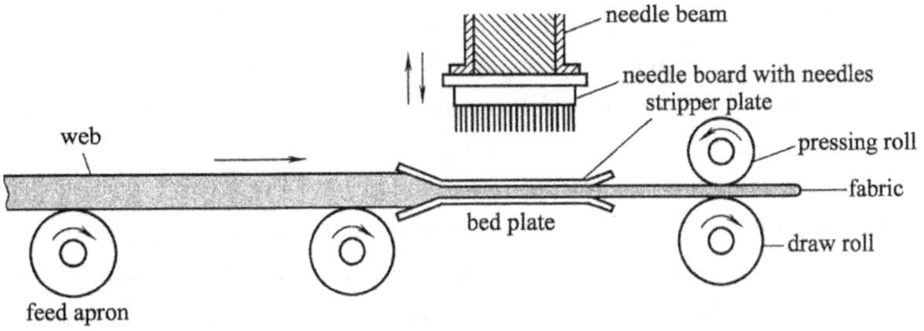

Fig. 6 – 6 Diagram of the needlepunching

There are several different types of needleloom. Fig. 6 – 6 is called a "down-punch" because it is pushing the fibers downwards. Similarly an "up-punch" pushes the fibers upwards. There is some advantage in combining an up-punch with a down-punch when trying to make a dense felt, rather than to punch continually in the same direction. For this reason some looms are made "double-punch", that is, one board punching down and one board punching up. Although this type of double punch loom is better than unidirectional punching, it has the disadvantage that all the down-punching is completed before the up-punching even starts.

下刺法
上刺法
双向刺法

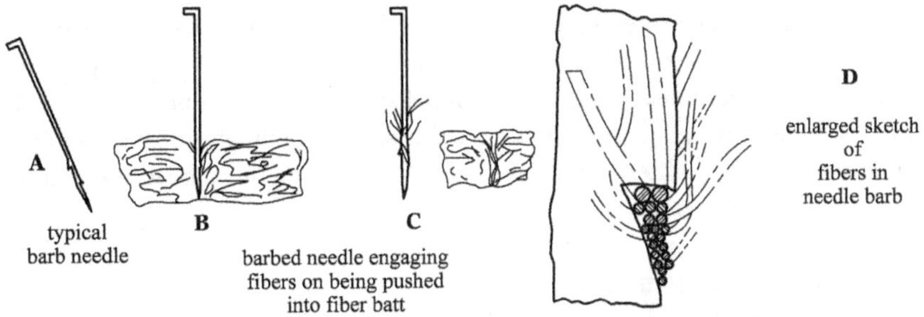

Fig. 6 – 7 Operation of the needles in the needlepunched process

Needlepunched nonwoven fabrics have a high breaking tenacity and also a high tear strength but the modulus is low and the recovery from extension is poor. For these last two reasons any needlepunching nonwoven fabric which is likely to be subjected to a load has to have some form of reinforcement, to control the extension. Needled carpets, for instance, may be impregnated with a chemical binder that gives better dimensional stability and increases the resistance to wear. In other cases thermal bonding may be used for the same object. For heavy duty applications such as filter media and papermakers' felts, yarns are used, either spun yarns or filament yarns. In some cases with a cross-laid web the threads may run only in the machine direction, but it is also common to use a woven fabric as the foundation and to build up the nonwoven on one or both sides.

针刺地毯
磨损
造纸毡

Needlepunched nonwoven fabrics are often known as needlefelts. They are used widely in gas filtration media and in some wet filtration. The principal advantage is that the nonwoven is practically homogeneous in comparison with a woven fabric so that the whole area of a nonwoven filter can be used for filtration. In papermakers' felts the same considerations apply, but in some cases the design of the felt has also to allow for the fact that the felt is acting as an enormous drive belt to drive some parts of the machine. In both gas filtration and papermakers' felts, needlefelts hold virtually the whole market. Needlefelts are used in geotextiles, but in view of the low modulus their application is mainly in removing water rather than as reinforcement. If the water is flowing through the fabric this is termed filtration. For instance a simple drain may be formed by lining a trench with a geotextile, partly filling it with graded gravel, drawing the edges of the geotextile together and backfilling the trench. The geotextile will keep the drain open for many years by filtering out fine particles.

排水沟/衬套/沟渠

砂砾

皮革

天然皮革/合成皮革/热收缩纤维

Many makers of synthetic leather have taken the view that the structure should be similar to natural leather. In these cases the backing or foundation of the synthetic leather is a needlefelt. The method of production is to include heat-shrinkable fibers which are made to shrink after intense needling in order to make the felt even more dense. After shrinking the felt is impregnated using a binder that fills the voids in the felt but does not adhere to the fibers, otherwise the felt would become very stiff. The leather is then finished by coating the face side with one or two layers of suitable polymer. Needlefelts are also widely used in home and commercial carpeting; in many cases the carpet may have either a velour or a loop pile surface to improve the appearance. Both velour and loop pile are produced in a separate needling operation using special needles and a special bedplate in the needleloom. Similar materials in a lighter weight are used for car carpeting, headliners and other decorative features on cars.

绒圈

6.3.4 Hydroentanglement 水刺法

Hydroentanglement by the use of high-pressure jets of water is another means of mechanically bonding webs. It was invented as a means of producing an entanglement similar to that made by a needleloom, but using a lighter weight web. A successful process was developed during the 1960s by DuPont and was patented. However, DuPont decided in the middle—1970s to dedicate the patents to the public domain, which resulted in a rush of new development work in the major industrial countries, Japan, USA, France, Germany and Britain.

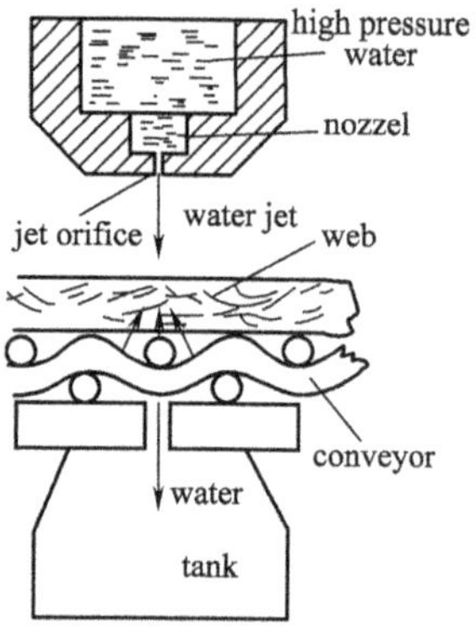

Fig. 6 – 8 Diagram of the hydroentanglement

The diagram of the hydroentanglement is shown in Fig. 6 – 8. As the name implies the process depends on jets of water working at very

水射流

high pressures through jet orifices with very small diameters. A very fine jet of this sort is liable to break up into droplets, particularly if there is any turbulence in the water passing through the orifice. If droplets are formed, the energy in the jet stream will still be roughly the same, but it will spread over a much larger area of web so that the energy per unit area will be much less. Consequently the design of the jet to avoid turbulence and to produce a needle-like stream of water is critical. The jets are arranged in banks and the web is passed continuously under the jets held up by a perforated screen which removes most of the water. Exactly what happens to the web underneath the jets is not known, but it is clear that fiber ends become twisted together or entangled by the turbulence in the water after it has hit the web. It is also known that the supporting screen is vital to the process; changing the screen with all other variables remaining constant will profoundly alter the fabric produced.

孔，口
水滴/紊流

Although the machines have higher through-puts compared with most bonding systems, and particularly compared with a needleloom, they are still very expensive and require a lot of power, which is also expensive. The other considerable problem lies in supplying clean water to the jets at the correct pH and temperature. Large quantities of water are needed, so recycling is necessary, but the water picks up air bubbles, bits of fiber and fiber lubricant or fiber finish in passing through the process and it is necessary to remove everything before recycling. The filtration process is more difficult than running the rest of the machine.

纤维润滑剂

Fabric uses include wipes, surgeons' gowns, disposable protective clothing and backing fabrics for coating. The wipes produced by hydroentanglement are guaranteed lint free, because it is argued that if a fiber is loose it will be washed away by the jetting process. It is interesting to note that the hydroentanglement process came into being as a process for entangling webs too light for a needleloom, but that the most recent developments are to use higher water pressures and to process heavier fabrics at the lower end of the needleloom range.

不掉毛

6.3.5 Stitch Bonding

缝编法

Stitch bonding is another kind of mechanical bonding. Webs can be consolidated by stitching either with a thread or without a thread. In the latter case the needles are designed to pull small loops of fiber out of the web and these are then used as the stitch material holding the web together. Stitch bonded fabrics have considerably stronger tensile strength in the direction of stitching than across the width.

缝编

Stitch bonding uses mainly cross-laid and air-laid webs. The web is taken into a modification of a warp knitting machine and passes between the needles and the guide bar(s). The needles are strengthened and are specially designed to penetrate the web on each cycle of the machine. The needles are of the compound type having a tongue controlled by a separate bar. After the needles pass through the web the needle hooks open and the guide bar laps thread into the hooks of the needles. As the needles withdraw again the needle hooks are closed by the tongues, the old loops knock over the needles and new loops are formed. In this way a form of warp knitting action is carried out with the overlaps on one side of the web and the underlaps on the other. Two structures are normally knitted on these machines, 链形缝编 pillar (or chain) stitch, or tricot. When knitting chain stitch the same guide laps round the same needle continuously, producing a large number of isolated chains of loops. When knitting tricot structure, 横移 the guide bar shogs one needle space to the left, then one to the right. Single-guide bar structure is called tricot, whereas the two-guide bar structure is often referred to as full tricot. The diagram of the stitch bonding is shown in Fig. 6 –9.

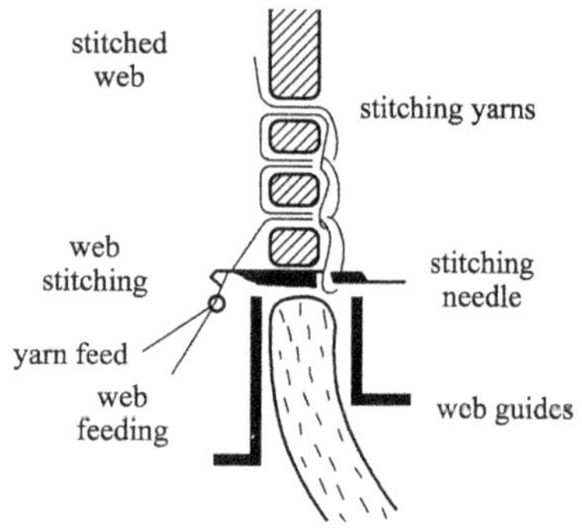

Fig. 6 –9 Diagram of the stitch bonding

The nature of this fabric is very textile-like, soft and flexible. At one time it was widely used for curtaining but is now 层合/床垫 used as a backing fabric for lamination, as covering material for mattresses and beds and as the fabric in training shoes. In deciding whether to use pillar or tricot stitch, both have a similar strength in the machine direction, but in the cross-direction the tricot stitch fabric is stronger, owing to the underlaps lying in that direction. A cross-laid web is stronger in that direction so the advantage is relatively small. The abrasion resistance is the same on the loop or overlap side, but on the underlap side the tricot fabric has significantly better resistance owing to the longer underlaps. However, continuous filament yarn is very expensive relative to the price of the web, so tricot fabric costs significantly more.

Stitch bonding can be processed without threads. In this case the machine is basically the same as in the above, but the guide bar(s) are not used. The needle bar moves backwards and forwards as before, pushing the needles through the web. The main difference is that the timing of the hook closing by the tongues is somewhat delayed, so that the hook of the needle picks up some of the fiber from the web. These fibers are formed into a loop on the first cycle and on subsequent cycles the newly formed loops are pulled through the previous loops, just as in normal knitting. The final structure is felt-like on one side and like a knitted fabric on the other. The fabric can be used for insulation and as a decorative fabric.

绒圈织物 To form a pile fabric by stitch bonding, two guide bars are usually used, two 绒圈纱/缝编纱/绒圈 types of warp yarns (pile yarn and sewing yarn) and also a set of pile sinkers, which are narrow strips of metal over which the pile yarn is passed and whose height determines the height of the pile. The pile yarn is not fed into the needle's hook so does not form a loop; it is held in place between the underlap of the sewing yarn and the web itself. It is clear that this is the most efficient way to treat the pile yarn, since any pile yarn in a loop is effectively wasted. This structure has been used for making towelling with single-sided pile and also for making loop-pile carpeting in Eastern Europe. The structure has not been popular in the West owing to competition with double-sided terry towelling and tufted carpets.

接结线 Laid yarns sewn together with binding threads is another structure of stitch bonding. Two distinct types of fabric can be made using the same principle. The first is a simulated woven fabric in which the cross-direction yarns are laid many at a time in a process a bit like cross laying. The machine direction yarns, if any are used, are simply unwound into the machine. These two sets of yarns are sewn together using chain stitch if there are only cross-direction threads and tricot stitch if machine direction threads are present the initial modulus of this fabric is very high compared with a woven fabric, which can first extend by straightening out the crimp in the yarn. These fabrics are in demand for making fiber-reinforced plastic using continuous filament glass and similar high modulus fibers or filaments.

The alternative system makes a multidirectional fabric. Again sets of yarns are cross-laid but in this case not in the cross direction, but at 45° or 60° to the cross direction. Two sets of yarns at +45° and −45° to the cross direction plus another layer of yarns in the machine direction can be sewn together in the usual way. Again high modulus yarns are used, with the advantage that the directional properties of the fabric can be designed to satisfy the stresses in the component being made.

Some examples of stitch bonded nonwoven fabrics are shown in Fig. 6 − 10.

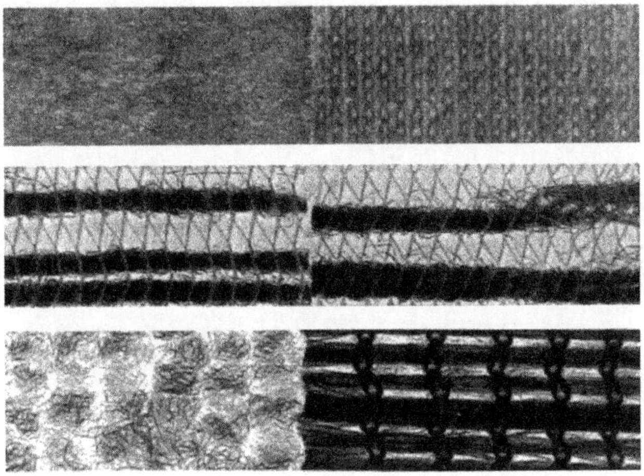

Fig. 6 − 10 Some examples of stitch bonded nonwoven

Review Questions

思考题

1. What are nonwoven fabrics?
2. What are the advantages of nonwoven over conventional fabric manufacture methods?
3. What are the main steps in nonwoven fabric manufacture?
4. Some web formations for making nonwovens are dry-laid and others are wet-laid. Describe the essential differences.
5. How many methods are there in dry-laying?
6. What are the problems in the parallel-laid? How can the problems be over-

come?

7. Please describe the basic principle of spun laying.

8. Give some examples of mechanical bonding.

9. How are needlepunched nonwovens different from bonded nonwoven fabrics?

10. What are the advantages of thermal bonding? Differentiate between melding and thermoplastic bonding.

11. Explain the basic principle of stitch bonding.

12. What limitations do nonwoven fabrics have? Indicate some uses of nonwoven fabrics.

13. Take some examples of nonwoven fabrics.

实训题 Suggested Activities

1. Collect examples of nonwoven fabrics. Determine fibers used and method of construction where observable.

2. Identify five trademarks of nonwovens and indicate the fibers from which they are composed.

Chapter 7 Other Fabric Construction Processes

织物的其他构成方法

Fabric construction processes discussed in this chapter include those with limited use and those that do not fit into the basic categories of weaving, knitting, and nonweaving.

7.1 Braiding

编织

Braiding is a method of interlacing (plaiting) three or more yarns or bias-cut strips of cloth over and under one another to form a flat or tubular fabric. These braided textile bands, which are relatively narrow, can be used as belts, pull-cords for lights, trimming for uniforms and dresses, tapes for pajamas, and some shoelaces. Several widths of plastic or straw braiding can be sewn together to make hat shapes. Similarly, braids of fabrics or yarns may be sewn together to make braided rugs. Strands are plaited together by crisscrossing them diagonally and lengthwise. A familiar illustration of the method is that of braiding long hair. Braid for fabric use is formed on a braiding machine by interlacing three or more strands of yarn so that each strand passes over and under one or more of the others.

斜裁

鞋带

Braids are divided into two types:

① Flat braids, in the form of strips or narrow flat tapes.

② Circular or round braids, tubular in form, which may be hollow or have a center core of some material (see Fig. 7-1 to Fig. 7-3). Flat braids are used for trimming. Circular braids appear in such everyday items as shoelaces and insulation for wires. All braided fabrics have considerable stretch in the length and some in the width. This property causes

平编织物

圆编织物

切边

(a) A 3-strand plait (the simplest form of braid)

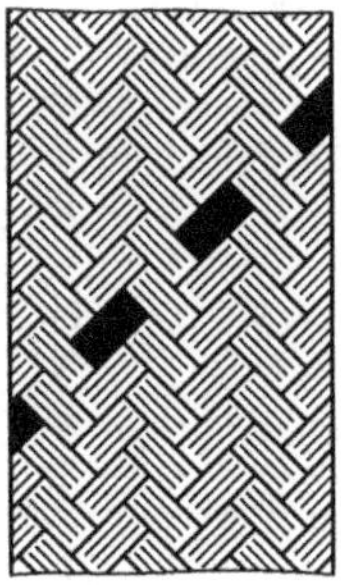

(b) Regular braid construction

Fig. 7-1 Diagrams of flat braids

编织装饰物 problems in the application of braided trim, because it is difficult to keep even. However, the stretchiness enables a skilled person to apply braid so that it lies smooth at corners and curved areas as well as on straight edges.

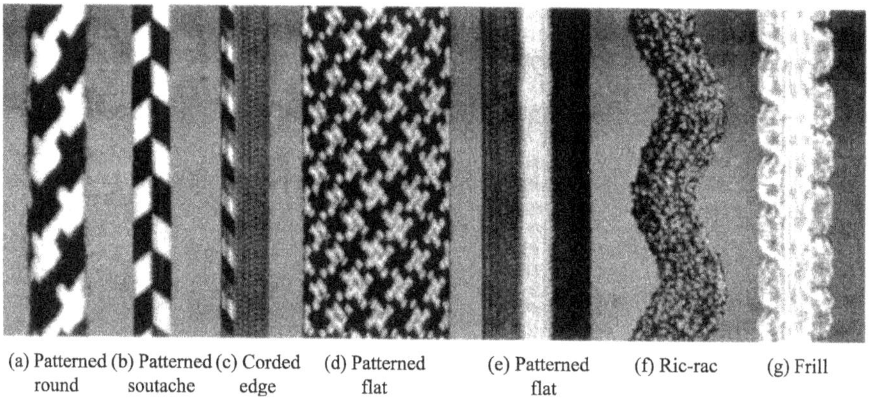

Fig. 7 – 2 Examples of fancy braids

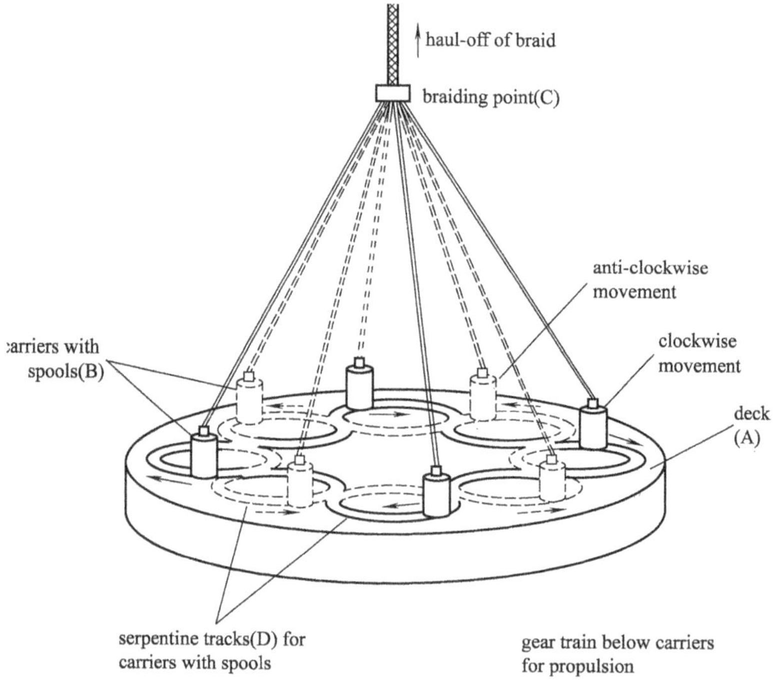

Fig. 7 – 3 Schematic of the basic principles of the round braider

Both types of braiding are produced from any of the textile fibers, as well as from metal threads, tinsel, straw, wire, or leather. This principle of fabric construction is used for making shaped articles, such as straw hats and small rugs; narrow fabrics, such as ribbons and braids for millinery and accessory dress materials; cords and tapes, such as fishlines, shoelaces, wicks, parachute and glider cords, and elastic of various types; and cord coverings for tires, tubing, hose, wires, and cables.

金属箔/稻草
女帽
灯芯/降落伞/滑翔带
轮胎/管状物/水龙带
电缆,绳索

Fabrics made by braiding are not limited to yarn forms. Cut strips of fabric, leaves, leather, straw, or any other flexible product can be braided to create attractive and unusual fabrics.

7.2 Netting (or knotting)

网织(打结)

Netting is an open-mesh form of fabric construction that is held together by knots or fused thermoplastic yarns at each point where the yarns cross one another. There are several types of mesh—square, hexagonal, and octagonal. The range of mesh sizes is from coarse and open to fine and shear. Netting may be made of any kind of fiber and may be given a soft or stiff sizing. Net fabrics are relatively fragile and require care in handling and cleaning. Torn net fabrics cannot be mended satisfactorily because the repair would be apparent. If the sizing is water soluble, the fabric should be dry-cleaned.

网眼

Nets are open-mesh fabrics. They have large geometric interstices between the yarns. Early nets were made by knotting the yarns at each point of intersection, and to some extent this is still done. Knotted nets have a comparatively large mesh and will not slip or spread. All net fabrics made before 1800 were done by hand. In 1809, a machine was developed that duplicated the net construction so accurately that only an expert could distinguish the handmade from the machine-made.

网状物

In recent years nets have been constructed on tricot and raschel knitting machines. They still have the open-mesh effect, but since knitting only interloops the yarns, these fabrics [Fig. 7 - 4 (a)] are not as durable as those formed by the knotting technique.

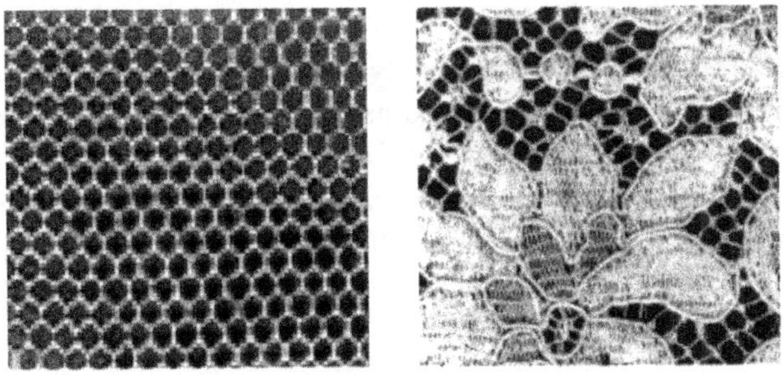

(a) Net fabric (b) Lace fabric

Fig. 7 - 4 Net and lace fabrics

Net construction is durable because of the knotting. The strength, combined

with the open areas, makes nets very effective in fishing, where water must drain away quickly, leaving the catch in the net. The same properties make nets very ef-
吊床 fective in various sports and for hammocks, curtains, and screens that must have good air circulation. The combination of durability and sheerness also makes nets
军用伪装网 attractive in evening clothes and millinery veils.

花边制作 7.3 Lace Making

Lace, which is a derivative of netting, has its origin in antiquity. In fact, the word for lace is derived from the Latin loquens, a noose or snare. Lace making is an art, the techniques of lace making involve looping, knotting, braiding, twisting, or stitching thread into decorative, openwork patterns. Lace was later used
新娘披纱 for bridal veils and other garment applications. As the fashion spread, a hand industry broadly developed among the European cities of Brussels, Lille, Mechlin, Valenciennes and Venice.

Lace has been defined in many ways, and authorities differ about what really
透孔织物 constitutes lace. However, it is safe to say that lace is an openwork fabric consisting of a network of threads or yarns formed into intricate designs [Fig. 7 – 4 (b)].

It is not known when lace was firstly made, but specimens in museums have been dated as early as 2500 B. C.. No other material is so difficult to make, yet so delicate, requires so much proficiency in manufacturing, or is so demanding of creative ability. Lace is a product of yarn twisting, and the machinery for lace making is among the most complicated known.

花边机 The development of lace making equipment involves some interesting history. Several early inventors devoted considerable time and energy to finding mechanical devices that would reproduce the intricate designs of handmade laces. In 1808,
机器编织的花边 John Heathcote completed a machine that successfully made bobbinet. This was followed in 1813 by the first Leavers machine. John Leavers and his brothers continued improving the machine and in 1837 incorporated the principle of the Jacquard loom attachment. The resulting product, the Leavers lace making machine, remains essentially the same today as it was in 1837.

Handmade laces have always been highly prized as trimming for apparel and as decorative pieces for the home. Unusually beautiful and intricately designed laces
传家宝 are retained in families as heirlooms and are displayed in museums as works of art.
针绣/织补/钩织 Today, such laces as needlepoint, bobbin (pillow), darned, crocheted, and
结网 knotted are duplicated so expertly by machine that the average consumer is not able to determine whether a lace is machine-made or handmade. Linen yarns had been generally used for expensive laces, but cotton, silk, rayon, and other yarns are now used for various qualities and types.

钩织花边 7.3.1 Crocheted Lace

Crocheting is a construction made from yarn with just one hook or needle, and a chain of loops is made from a single yarn. Crocheted lace is made with a hook to form a series of loops, each one of which is finished with a fine stitch, It originated in Ireland as an imitation of Venetian needlepoint.

7.3.2 Tatting Lace 梭织花边

Knotting (or netting) is the process of forming an open-work fabric or net by tying yarns together where they cross one another as in lace. Tatting lace is a knotted lace made with the aid of a small shuttle filled with yarn. Clover leaf and wheel designs are the most popular, but other patterns are also made.

7.3.3 Macrame Lace 流苏花边

Macrame differs from other laces in texture and appearance. It is generally made of heavy yarn knotted into relatively large designs (see Fig. 7-5). Unlike most other laces, it is used for such purposes as women's hats, handbags, belts, and vests (see Fig. 7-6). 背心

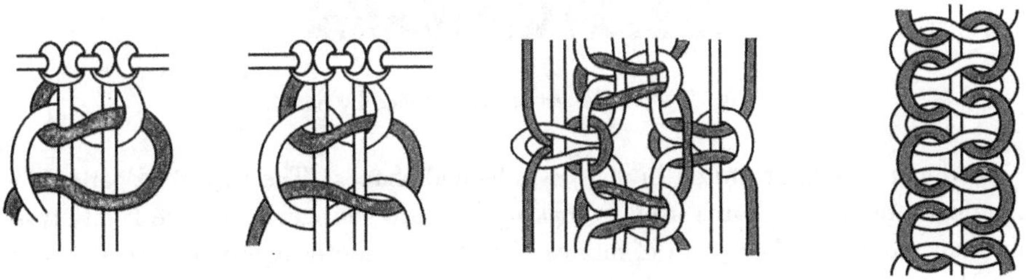

(a) Square knot (b) Alternating square knot pattern (c) Square knot sinnet

Fig. 7-5 Basic Macrame knots and some variations

Fig. 7-6 Macrame lace of knotted heavy yarn may be made into such garments as this vest

7.3.4 Bobbin (Pillow) Lace 纱管花边

A pattern for bobbin lace is drawn on paper which is then placed on a pillow and held with pins set into it in conformity with the design. A number of threads, each on its own bobbin, are interlaced by twisting and plaiting around the pins to produce the motif in a mesh construction (see Fig. 7-7). There are several varieties of bobbin lace.

花纹图案/网目

Fig. 7-7 The art of making lace by hand

Lace can be manufactured in many widths and shapes. The variety of patterns is practically limitless. Yarns used in making lace are stronger and more firmly twisted than those for other types of fabrics. The adaptation of man-made fibers (such as nylon, with its inherent high strength) has added durability to lace.

层合 7.4 Laminating

衬背

湿法黏合织物/泡沫材料黏合织物

Laminating is the process of joining a fabric to a plastic foam backing or bonding two fabrics together. There are following two techniques used to produce laminated fabric. Wet adhesive bonded fabrics normally have better quality than foam bonded fabrics.

湿法黏合

(1) Wet Adhesive Bonding

丙烯酸

干洗

In this case a water soluble acrylic base adhesive is coated on the back of the face fabric which is then joined to the liner. The fabric is then passed through a first stage of heating which expels the solvents and creates a preliminary bond known as "green" tack. The fabric then passes through a second heating stage which creates a strong permanent bond to stand washing and dry cleaning.

泡沫塑料热熔黏合

(2) Foam-Flame Bonding

聚氨酯泡沫材料

The adhesive used in this process is polyurethane foam. By applying a gas flame to the foam, it is burned off and becomes tacky on one side so that it then adheres to the face or first fabric as passage occurs through the rollers.

The combination is then cooled and flame is applied to the other side of the foam causing it to become tacky and adhere to a second, liner, fabric. The tension control and the flame control are two main factors to be carefully considered. The method is sometimes used to provide a backing for light weight fabrics.

Fabric composites have become widely used in many forms, fabric to fabric bonding is used to achieve good properties by using two low quality fabrics. Stability can be given to light weight, open construction woven or knitted fabrics by bonding them to a warp knit tricot fabric. In this way, bonding creates a unique group of textiles which cannot be produced by any other method economically. The two previous systems are not usually considered fabric forming systems, but as finishing processes.

7.4.1 Laminated Fabrics

层合织物

A laminated fabric is one in which at least two layers of material or fabric are sealed together. Adhesives may be used such as thermosetting resins, thermoplastic substances, heat treated polyurethane foam.

热固性树脂/热塑性物质

One of the most common laminated fabric is made of two fabrics securely bonded or sealed together by an adhesive [see Fig. 7-8(a), Fig. 7-9]. These structures frequently use tricot knit as the backing fabric, in order to provide a self-lining and to stabilize the fabric. Bonded fabrics have good stability, resistance to stretch or deformation, and a low incidence of raveling.

自衬背

脱散

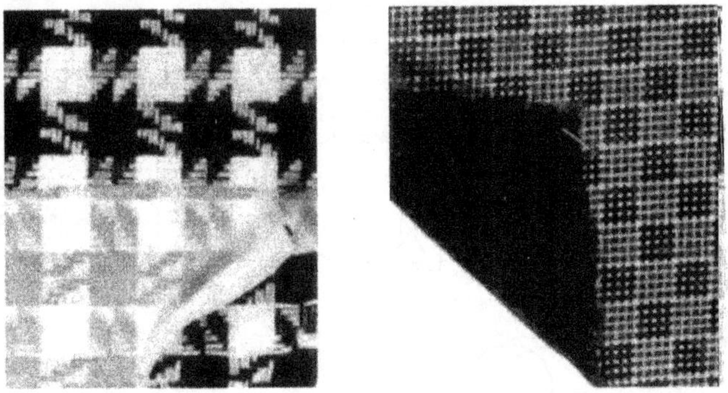

(a) Bonded fabric-to-fabric (b) Bonded fabric-to-foam

Fig. 7-8 Laminated fabrics

Another type of laminated fabric has a layer of fabric bonded to a layer of foam, which adds warmth and stability [Fig. 7-8(b)].

In sandwich construction a layer of fabric is bonded to a second fabric by means of a thin layer of foam in the center (Fig. 7-10). Besides providing stability and warmth, the sandwich laminate has a finished appearance on both sides.

7.4.2 Problems of Serviceability of Laminated Fabrics

层合织物的使用问题

Layered fabrics sometimes come apart (become delaminated) by abrasion, because the adhesive cracks and the face fabric separates from the backing, or because in laundering one fabric shrinks more than the other due to poor stabilization. In dry cleaning, the fabric layers pull apart if the solvent solubilizes the adhesive. Designs formed by spot welding discolor or disappear—a common occurrence in simulated quilted fabrics. Mechanical action in dry cleaning might cause separation of the fabric and obliteration of the design. However, there has been a

多层织物

洗涤

绗缝织物

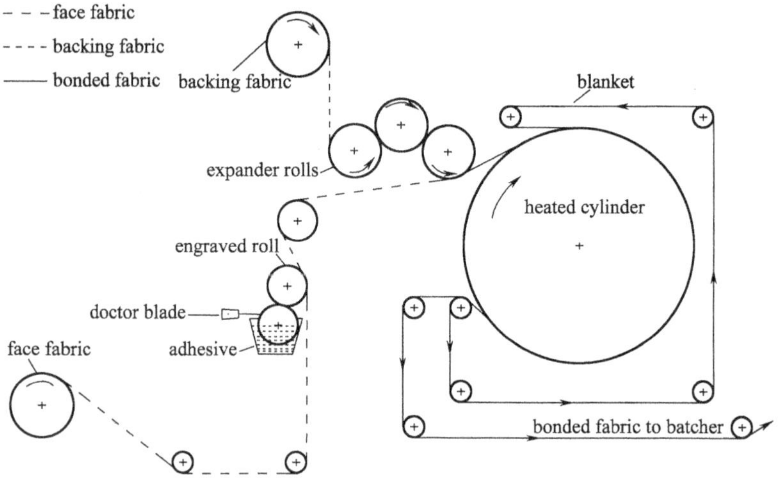

Fig. 7 – 9 Diagram of the fabric-to-fabric bonding process

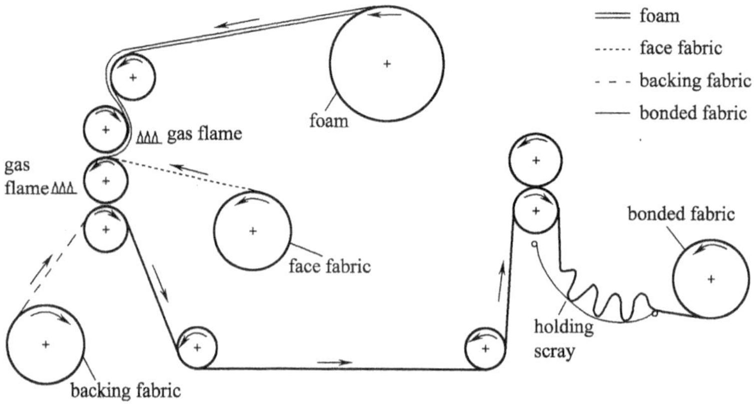

Fig. 7 – 10 Diagram of the fabric-to-foam-to-fabric bonding process

decided improvement in resistance to delamination.

Urethane backing can attract loose soil or dye particles during dry cleaning or use. This often occurs with foam-backed (泡沫塑料衬里) mats. The foam turns yellow or darkens owing to exposure to heat and atmospheric conditions. The outer layer of a garment can change color. Ripples and puckers (波纹/起拱) might be caused by a shrinkage of one of the layers of the cloth, The spunbonded fabrics might have finishes that are not colorfast (色牢度强的) to light, washing, and perspiration. Again, the foam backing applied to upholstery may not be resistant to solvents in cleaning, even though the material itself is highly resistant. Testing with a dry-cleaning solvent will determine whether gentle washing should be substituted.

The consumer should examine laminated fabrics to determine the following points:

① The fabric layers are sealed together firmly.
② The grain lines are held in their proper relationship.
③ Adequate care instructions are included.
④ There is no unusual odor.
⑤ There is no stiffness.
⑥ The adhesive is not visible.
⑦ The foam, if used, is not discolored.
⑧ The stitching on quilted fabrics uses small stitches, and there are no breaks.

7.5 Tufting 簇绒

Tufting is a method by means of which looped piles or cut piles are inserted into a base fabric. The process was first utilized only in the manufacture of carpets, but tufted fabrics (簇绒织物) have been used in many other areas. Unlike woven pile fabrics, the pile yarns in tufted fabrics do not interlace with the ground yarns. In manufacture, the pile yarns, threaded in needles, penetrate the backing cloth. On the face side the yarn is held by a looper which prevents it from being withdrawn when the needle recedes as shown in Fig. 7 – 11.

See Fig. 7 – 11 (a), each time the cloth is fed forward, the tufting needle pushes the yarn through the cloth to form a loop pile. The pile hook moves forward and catches the loop as the knife moves with it to cut progressively the rearmost loop on the hook.

See Fig. 7 – 11 (b), the needle inserts the yarn through the fabric to form a loop pile. The hook grasps the loop, holding it in place as the needle is retracted. The loop height depends upon the rate of yarn feed, and can be determined and adjusted by the combined depth of needle thrust and hook placement level.

Many types of yarns can be used for the pile such as wool, cotton, viscose, acetate, nylon or polyester. The backing cloth must be sufficiently firm, otherwise it might become distorted during processing. Jute was the main fiber used to produce woven fabrics for backing cloth. Recently there has been a trend to use other fibers such as polypropylene and to use nonwoven fabrics. The backing cloth must hold the yarn loops firmly enough during processing, but the permanent anchoring (锚定) is achieved by applying a layer of adhesive on the back side of the tufted fabric. In many cases a secondary backing is also used.

By controlling the amount of yarn being fed to each needle, the size of the loop is determined, and this in turn produces variations in texture and design. Tufts can be cut or uncut, and combinations of both are used to create interesting effects.

Color depends on the arrangement of colored yarns, the use of fiber variants that accept different types of dyes, or printing patterns onto the surface. Other design effects can be achieved by the spacing and height of the tufts.

Identification of tufted fabrics is comparatively simple if there is no secondary backing. Tufts run in the lengthwise direction, and the underside shows the back loop of the tufts as a separate yarn, not an integral part of the weave (Fig. 7 – 12).

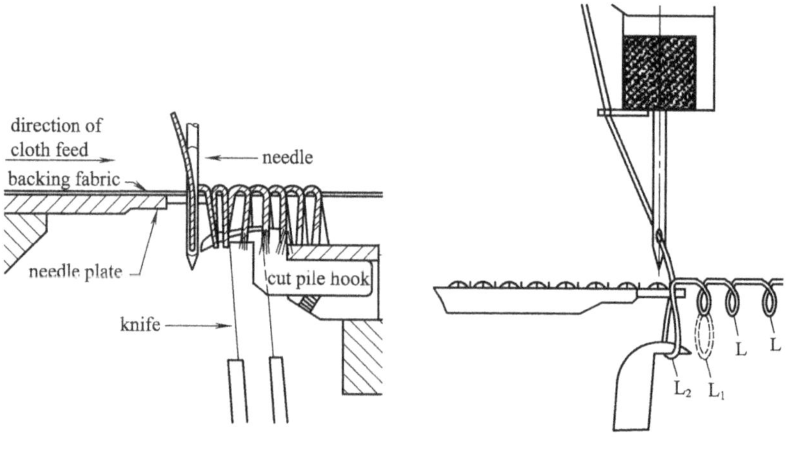

(a) Cut pile (b) Loop pile

Fig. 7-11 Tufting machine

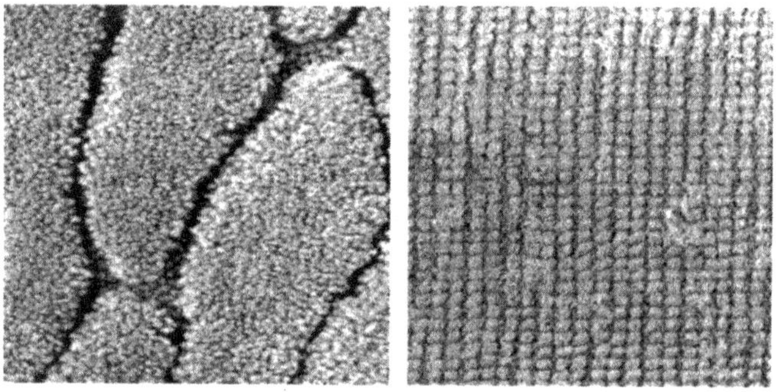

(a) Face (b) Back

Fig. 7-12 Tufted carpet. Backing fabric, usually added for stability, has been removed to show the tufting yarns on the back of the ground fabric

簇绒地毯

7.5.1 Tufted Carpets

Tufted carpets and rugs made their appearance in 1950 with the development of special machinery. The success and popularity of these carpets has been phenomenal. Approximately 90 percent of all carpets produced in the United States are made by the tufting process. This predominance of tufted carpets derives from the fact that a great variety of fibers can be used successfully, and durable carpets can be made for lower cost than by other construction methods. It can achieve a wide variety of textures and patterns. It is fast and can produce carpet that is 360 cm wide at the rate of about 190 cm in length per minute.

Tufted carpets are composed of pile yarn inserted into a base fabric called a primary backing which is usually a woven fabric made of jute or, more frequently, polypropylene. The primary backing may also be a nonwoven fabric. 基布,底布

To secure the tufts permanently in place, most tufted carpeting is either sealed to another fabric to increase dimensional stability and prevent damage or is coated with a heavy layer of adhesive or sealing compound (see Fig. 7 – 13).

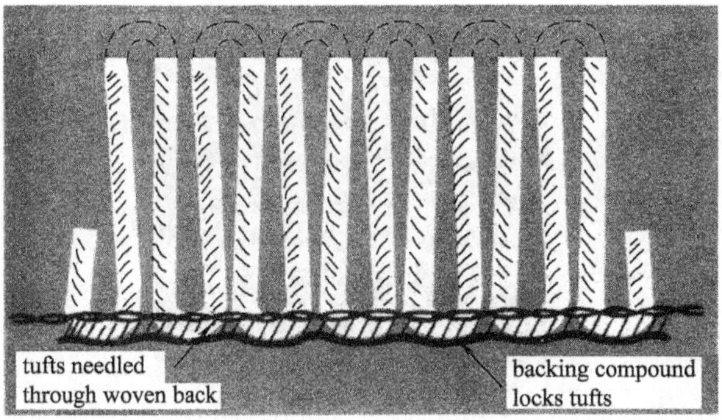

Fig. 7 – 13 Tufted construction

7.5.2 Types of Tufted Carpets 簇绒地毯类型

The nature of the pile—its twist and height, whether it is cut or uncut, and its density—determines the surface appearance and texture of tufted carpet. The surfaces are identified according to type, although there is some overlapping of terminology.

7.5.2.1 Shag 粗毛

Shag carpets are made with relatively coarse cut or a combination of cut and uncut twisted pile that is generally an inch (25.4mm) or longer. Because of its height, the pile tends to lie in a somewhat random fashion rather than stand erect. It tends to crush but can be somewhat restored with a special brush or rake that raises the pile (see Fig. 7 – 14). Shag carpet of continuous filament nylon pile tufted through a web of nylon staple needlepunched to a secondary polypropylene backing. The primary nylon web backing of the same fiber and color of the pile gives the carpet the appearance of a greater pile density than there actually is. 粗毛地毯 耙子

Fig. 7 – 14 Shag carpet

平铺绒圈 **7.5.2.2 Level Loop**

As the term implies, level loop is an uncut pile. It is of uniform low to medium height. The pile generally has a moderate twist. Density varies see Fig. 7 - 15 (a).

割毛与不割毛联合 **7.5.2.3 Combination Cut and Uncut Pile**

Attractive surface-texture effects are obtained by combining cut and uncut pile, sometimes in interesting patterns [see Fig. 7 - 15(b)].

(a) Level loop

(b) Cut and uncut pile

(c) Variations in pile height

Fig. 7 - 15 Tufted carpets

浮雕，刻纹 **7.5.2.4 Sculptured**

The pile may be woven in different heights to form either a definite or a random pattern, which provides uneven but interesting and attractive patterns, see Fig. 7 - 16(c). This sculptured effect is also referred to as carved or multilevel. Combination cut and uncut pile may be utilized in this construction.

浮雕效应/雕刻/多个凹凸

思考题 **Review Questions**

1. At least three strands required in braiding, why?
2. What braided articles are used in apparel or in home furnishings?
3. Why is braiding confined to the production of shaped articles?
4. What does net differ from woven fabrics and lace?
5. Compare the methods of making the following laces: bobbin, crocheted, and tatting.

6. What is Macrame Lace and its characteristics ?

7. What does the quality of lace depend upon? Describe the uses and the cares for lace.

8. How are fabrics laminated? For what purposes are laminated fabrics used?

9. Explain the merits and drawbacks of laminated fabrics in use.

10. Introducing two types of tufting processing.

11. Why have tufted carpets become the predominant type in United States?

12. Describe at least three types of tufted carpet face surfaces.

Suggested Activities 实训题

1. Visit the yard-goods and lace departments of your local store. List in separate columns all fabrics that are woven, knitted, nonwoven, net or lace, braided, laminated, or other construction. What percentage of the fabrics listed comes in each category? What conclusions can you draw?

2. Collect examples of other construction fabrics mentioned in text of this chapter. Determine fibers used and method of construction where observable.

Chapter 8 Textile Dyeing and Finishing

纺织品染整

前处理 8.1 Preliminary Treatments

Most textile materials and fabrics require preliminary treatments before they can be dyed and finished. The reason why such treatments are necessary is to make sure that the dyeing, printing and finishing are acceptable, predictable and reproducible. The impurities present, either natural or man-introduced, must be removed. These preliminary treatments are approximately the same for all cloths, but they vary to some extent with fiber type. They include singeing, desizing, scouring, degumming of silk, bleaching and occasionally mercerization of cotton. Each of these processes is discussed in the following sections.

烧毛/退浆/煮练

烧毛与退浆 8.1.1 Singeing and Desizing

Singeing is the first operation of the preparatory processes necessary for dyeing or printing. Singeing removes protruding ends of fibers in order to produce a clear, smooth, and uniform surface. Singing is usually done by passing the fabric through a burning gas flame at high speed followed by quenching in water or the desizing bath to extinguish any sparks.

退浆浴

Singeing is sometimes done after scouring since heating of the fabric in the singer can increase the difficulty of removing size and soil from the fabric. Singeing and desizing are frequently combined. When combined with singeing it prevents possible latent damage to fabric from the singeing flames or heated plates.

烧毛机

Desizing removes size from fabrics. Sizing materials are applied to yarns, particularly warp yarns, before they are woven into cloth. These form a protective coating over the yarns and keep them from chafing or breaking during weaving. Most of the size must be removed before the fabric can be dyed satisfactorily. Residual size prevents the yarns and fibers from wetting quickly and can affect dye absorption in either batch or continuous dyeing.

摩擦

匹染

煮练与漂白 8.1.2 Scouring and Bleaching

果胶

Scouring is an operation in which any pectin, ash, dirt, wax or other substances that may have adhered to the fibers in the processing of the yarns is reduced to an amount which will not seriously interfere with subsequent dyeing. Bleaching follows scouring in the continuous scouring and bleaching system. Bleaching is a final treatment to complete the purification, which at the same time ensures the production of a good white color and increased the ability of the fabric to absorb dyestuffs uniformly if it is to be dyed or printed. 净化

Soaps or synthetic detergents with alkali builders are common scouring agents. For protein fibers a neutral or slightly acidic synthetic detergent is often used. Despite the use of chemicals, no molecular change occurs in the fiber. During the scouring operation chemicals applied during desizing that have remained in the fabric are removed. 肥皂/增效剂 中性的/弱酸性的

The particular chemical used for bleaching depends on the textile fiber. Cellulosic fibers, such as cotton, linen and the various rayons, can be bleached with chlorine compounds(sodium hypochlorite is one of the chemicals used) or hydrogen peroxide. Perborate bleaches are available for home use, but they are seldom employed commercially since hydrogen peroxide reacts in much the same way and is more effective. 氯化物/次氯酸钠 过氧化氢/过硼酸盐

Fabrics may be bleached in rope or open-width form. In rope bleaching the fabric is pulled together to form a somewhat circular mass, which is loose enough for penetration and resembles a large rope; in open-width form the fabric is under tension and is flat and smooth. 绳状/平幅

Typical steps in scouring and bleaching for either cotton or cotton blend fabrics in rope form are as follows: Fabric is fed to a prewasher, caustic saturator, J-box, washers, peroxide saturator, J-box, washers, and finally to storage bins (Fig. 8 - 1). 预洗机/浸碱池/J形箱/储布池

Open-width bleaching differs in minor ways. The fabric is handled in open flat width under tension to prevent formation of crease marks(Fig. 8 - 2). Instead of passing through J-boxes, the fabric passes through steamers on a series of rollers. Fabric moves through the unit at 80 - 120 yards(73 - 110 m) per minute. 折痕 蒸箱 码

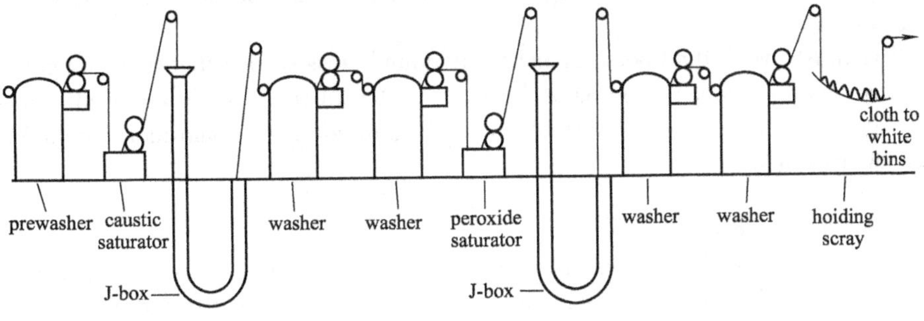

Fig. 8 - 1 Diagram showing basic steps in rope scouring and bleaching

An additional technique available for bleaching utilizes optical brighteners. 荧光增白剂

A special scouring process used to remove the natural gum, sericin, from silk, is called degumming. The silk is washed in hot soapy water, usually under alkaline conditions. Sericin functions as a size in the manufacturing process and is generally removed after the yarn is woven or knitted.

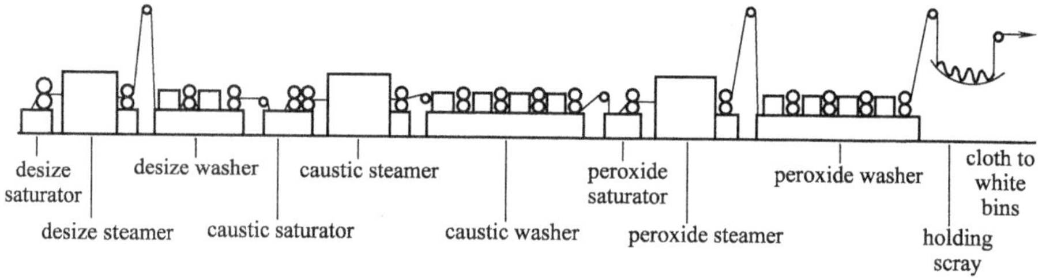

Fig. 8-2 Diagram showing basic steps in open-width bleaching and scouring

丝光 8.1.3 Mercerization

Mercerization is a chemical finish applied to cellulosic fibers, especially cotton. It adds luster to fabric, improves dyeing characteristics, and increases strength. Steps in mercerizing are wetting out (浸湿) the fabric with water; saturating with caustic; timing to permit mercerizing action; tentering (拉幅) to set fabric dimensions; washing; neutralizing; washing and drying (Fig. 8-3).

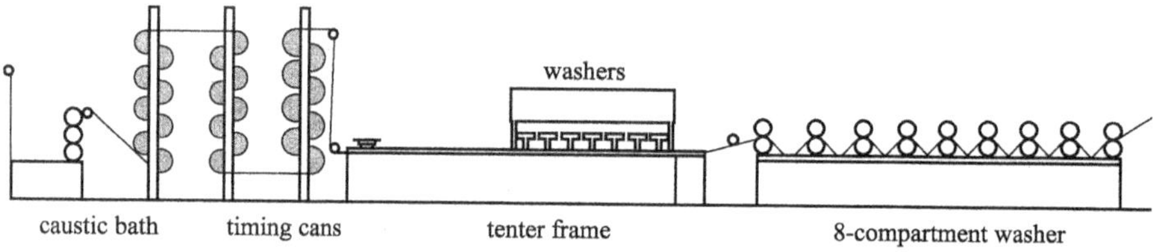

Fig. 8-3 Diagram showing basic steps in mercerizing

Fabric moves through the mercerizing unit (丝光机) at 80-120 yards (73-110 m) per minute, depending on fabric weight and compactness and on the concentration of the NaOH.

The finish swells (溶涨) the fibers, giving them around cross section that reflects light to create a gloss or sheen (光泽). The natural twist of cotton fiber is largely removed. Mercerization under tension produces fibers with increased strength and increased affinity (亲和力) for dyestuffs.

热定形 8.1.4 Heat Setting

Thermoplastic or heat-sensitive fibers (热塑性或热敏性纤维) are generally given a heat-setting finish to produce a special shape or to ensure a stable fiber. Both filament and staple fiber fabrics composed of heat-sensitive fiber types are heat set. The process is physical and alters the physical characteristics of the fiber.

These fibers have both a melting temperature and a glass transition temperature (玻璃化转变温度) (T_g). The glass transition temperature is the point at which the amorphous regions (无定形区) of the fiber develop flow or "melt". At this temperature fibers can be shaped. Only the interior of the fiber need reach the T_g temperature.

In addition to the flow of the amorphous regions, there are some changes in the crystalline areas of the polymer. Small crystallites may disappear and large crystallites may grow larger and assume the configuration in which the fiber is held smooth, bent, folded, or uneven.

结晶区
构型

The rules for heat setting are followed:

① Fiber temperature must reach or exceed the T_g before the fiber is allowed to cool.

② Any subsequent resetting or reshaping must be done at a temperature greater than that used in the first setting.

A major reason for heat setting is to introduce dimensional stability, and the degree of dimensional stability is determined by the temperature, the period of exposure, and the amount of force used to hold the fabric in the desired shape and size during the setting. Other characteristics introduced by heat setting include resiliency, which contributes to wrinkle resistance; elastic recovery, which aids in size retention; and relatively permanent design details. Fabrics can be heat-set in a smooth, flat shape or with pleats and creases pressed in.

褶皱/皱痕

8.2 Dyeing and Printing

染色与印花

Color can pleasure us. The pleasure derived from imparting color to clothing has existed since the time of the earliest civilization. Life would be more hazardous and certainly dull if textiles were only used in their natural colors.

Textile's color is normally obtained by applying a colorant to the textile substrate. There are two ways of adding color to a textile substrate: i. e. dyeing and printing. Printing adds color to the substrate locally; whereas dyeing completely covers the substrate with color, and usually with the intension of obtaining an even color distribution throughout the substrate. Colorants used for dyeing can be classified as dyes or pigments. Dyes are water soluble and have substantively affinity for fibers and can be absorbed into the fibers. Pigments are not water soluble and possess no specific attraction for any particular fiber type. Pigments usually adhere to the surface of the fiber. Dyes are used far more widely and frequently than pigments in the textile dyeing process.

着色剂
被染物，基质
涂料

8.2.1 Dye Classes by Application Method

染料按应用方法分类

The technical definition for a dye is a compound that can be fixed on a substance in a more or less permanent state and evokes the visual sensation of a specific color. In order to research and develop and use them more conveniently, dyes are usually grouped in a number or ways depending upon their ionic character, chemical type or application method (see Table 8 - 1). However, grouping dyes by their application methods is common used because it is most useful to the technologist concerned with coloration of textile products.

离子的

Based on their application methods, dyes are essentially divided into nine types, which are substantive or direct dyes, acid dyes and mordant or metallized dyes, azoic or naphthol dyes, cationic or basic dyes, vat dyes, sulfur dyes, reactive dyes, disperse dyes and pigments. Next table summaries the scope for use and properties of some of the major application classes of dyes.

媒染剂/金属络合染料/不溶性偶氮染料/阳离子/活性染料/分散染料

Table 8 - 1 Summary of Major Dye Types

Type	General description	Uses	Comments
Direct	Dye cellulosics directly; some dye wool, silk, nylons; blends; complete shade range.	Cellulosics cotton, viscose.	Colors duller than basic or acid dyes. Cheap, poor washing fastness.
Acid	Originated from basic dye acidification; complete color range.	Primarily for wool and silk; also acetate, nylons, acrylics, spandex;	Has good leveling; poor washing fastness.
Mordant	Related to acid dyes; require addition of chrome derived from potassium or sodium bichromate .	Primarily for wool requiring maximum fastness; carpeting, men's wear; silk and cellulose fibers.	Colors duller than acid dyes. Good fastness to light and washing.
Azoic	Also known as insoluble azos and ice colors; ice helps to reduce temperature of dyeing bath.	Primarily for cotton goods; extensive printing use since colors are dischargeable.	Yellows, reds, blacks most used; good fastness to light and washing.
Basic	First synthetic dye (1856); organic base dissolved in inorganic acid; limited use today; complete color range.	Cotton, wool, silk, nylon, polyesters, acrylics; direct printing on acetate; discharge printing on cotton.	Bright shades, excellent tinctorial strength.
Vat	Synthetic indigo original; insoluble in water; require reduction to apply.	Primarily for cotton, also wool; generally for work clothes, outerwear.	Expensive; bright colors often difficult to achieve.
Sulfur	First created in 1879; generally insoluble in water; complete shade range except for true red.	Applicable to stock, yarn, piece goods; used for heavy woven and knitted cotton and linen.	Colors not bright.
Reactive	First available in 1957; form chemical combination with fiber, distinguishing class from others.	Primarily for cotton apparel, decorative fabrics for bright colors; some suitable for wool, silk and nylons.	Produce brightest shades; excellent shade range, high fastness due to covalent dye/fiber bond.
Disperse	Insoluble in water; supplied in paste or finely powdered from; particles disperse in water and dissolve in fibers.	All synthetic fibers, wide use in apparel, decorative fabrics; used for dyeing and printing.	Good shade range; best fastness on polyester, good leveling properties.
Pigment	Organic coloring materials; insoluble in water with no affinity for fiber; fixed on fiber with resinous binders.	Primarily for printing cotton of all weights; also wool, rayon, acetate, nylons, polyesters, olefins.	Complete shade range in bright colors; good fastness to washing and light.

牢度

酸化/匀染性

铬
重铬酸钾或重铬酸钠

拔染印花/着色强度

色谱全
共价的

树脂黏合剂

8.2.2 Methods of Dyeing 染色方法

Textiles may be dyed at any stage of their development from fiber into fabric or certain garments by the following methods:

① Stock dyeing, in the fiber stage. 散纤维染色

② Yarn dyeing, after the fiber has been spun into yarn.

③ Piece dyeing, after the yarn has been constructed into fabric. 匹染

④ Solution or dope dyeing, before a man-made fiber is extruded through the spinnerette. 纺前染色

⑤ Garment dyeing, after certain kinds of apparel are knitted.

8.2.2.1 Stock Dyeing 散纤维染色

Stock dyeing refers to dyeing a staple fiber before it is spun. There are two methods. The older and widely practiced procedure is that of removing the packed fiber from the bales and then packing the stock in large vats and circulating dye liquor through the mass of fiber at elevated temperatures [see Fig. 8-4(a)]. The newer method, bale dyeing, which is applicable to wool and all types of man-made fibers, is that of splitting the bale covering on all six sides, placing the entire bale in a specially designed machine the covering and straps need not be removed, and then forcing the dyed liquor through the bale of fiber [see Fig. 8-4(b)]. This latter method obviously saves time and labor costs.

染缸

原坯冷染

分开/覆盖物

(a) Stocking dyeing (b) Bale dyeing

Fig. 8-4 Devices of several dyeing methods

Although the dye liquor is pumped through the fiber in large quantities, there may be areas where the dye does not penetrate completely. However, in subsequent blending and spinning operations, these areas are so mixed with the thoroughly dyed fiber that an overall even color is obtained. In stock dyeing, which is the most effective and expensive method of dyeing, the color is well penetrated into the fibers and does not crock readily. Stock-dyed fiber does not spin as readily as undyed fiber because it loses some of its flexibility, but lubricants added in the final rinsing overcome most of this difficulty. 柔韧性

Woolens are often stock-dyed. The completeness of this method is reflected in the expression "dyed in the wool", which is used to attribute the quality of thorou- 粗纺毛织品 彻底性

牛津布/粗花呢，手工纺织呢 ghness. Stock dyeing produces mixture effects and color blends, of which oxford suitings and tweed homespuns are examples.

纱线染色 8.2.2.2 Yarn Dyeing

When dyeing is done after the fiber has been spun into yarn, it is described as yarn dyeing. There are several methods of yarn dyeing. The purpose is to have the dyestuff penetrate to the fibers in the core of the yarn; this is similar to the penetration of the fibers in stock dyeing. Cloth made of dyed yarns is called yarn-dyed.

Yarn-dyed fabrics are usually deeper and richer in color. Yarn-dyed fabrics intended for laundering must be quite colorfast, or bleeding could occur. The prima- 条纹 ry reason for dyeing in the yarn form is to create interesting checks, stripes, and 格子花纹/条纹布 plaids with different-colored yarns in the weaving process. Chambrays, for example, are usually woven with a colored warp and white filling. Other combinations of 绉条纹薄织物/混色毛纱 different-colored yarns are checked gingham, shepherd's check, plaid, seersucker, and heather mixtures.

匹染 8.2.2.3 Piece Dyeing

单色 Most solid-color fabrics are dyed after the fabric has been completed. This is piece dyeing, the easiest and cheapest method for adding color. Manufacturers can 库存 color fabrics as ordered and need not maintain a large stock of dyed fabrics that might become obsolete if the fashion changes. Piece dyeing does not always provide a thorough penetration of dyestuff, but for many uses it is quite satisfactory. Normally, piece dyed fabrics are a single color, but when a fabric contains more than one fiber, a pattern can result from the different absorption rates of the fiber and different levels of attraction or reaction between dyes and fibers. The most im- 混染/交染 portant variations are union dyeing and cross dyeing.

混染 (1) Union Dyeing

In union dyeing, a fabric containing two or more fibers is dyed a single uniform color. Various dyestuffs are applicable only on certain fiber types, so when a fabric containing two or more types of fibers is to be colored a solid color, the dyes must be carefully selected and properly applied in order to ensure color uniformity. If possible, dyes should be chosen that can be applied simultaneously to the fabric. However, in some cases dyers have found it preferable to introduce the dyes individually, each by its recommended procedure. These two methods are often called 一浴法/两浴法 one-bath and two-bath processes.

交染 (2) Cross Dyeing

In cross dyeing, a fabric containing two or more fibers is dyed so that each fiber accepts a different dyestuff and becomes a different color. In some cases the dye bath is planned so that certain fibers will accept no color at all and remain white. The end product depends upon the fiber arrangement in the fabric. It may be a check, a plaid, a twill, a stripe, a muted color, or some other design. Recent developments in fiber chemistry have produced fibers of the same generic type that 染浴 are dye selective, so that fabrics of one fiber come from the dyebath checked or striped. This is especially successful with acrylic and polyester fibers.

8.2.2.4 Solution or Dope Dyeing

原液染色或纺前染色

In the manufacture of man-made fiber, color can be added to the chemical solution before it is forced through the spinnerettes. If the pigment is good, this method ensures not only even dyeing but colors that are an integral part of the fiber and, therefore, fast to most outside influences. This solution dyeing or dope dyeing is especially important in coloring acetate and olefine fibers.

坚牢

8.2.2.5 Garment Dyeing

成衣染色

Certain kinds of nontailored apparel, such as hosiery, pantyhose, and sweater can be dyed as completed garments because they are each made of a single component and will not be readily distorted. However, allowance must be made for anticipated shrinkage. A number of garments are loosely packed into a large nylon net bag. The bags are then put into a paddle dyer, which is a tub with a motor-driven paddle that agitates the dye bath.

袜裤

浆叶式染色机/槽/电动机驱动的

Garment dyeing is an economical method and is used when practical. It also reduces the risk of building an inventory that could be affected by changes in color fashion.

库存

8.2.3 Printing

印花

Printing produces localized coloration of textile materials and is often used to produce colored patterns on fabrics or garments. Each color applied in a printing process must be applied in a separate step or position in the printing machine. Printing methods may be classified as follows: direct printing, discharge printing and restist printing.

局部染色

直接印花/拔染印花

防染印花

Printing methods may also be classified according to the process used to produce the pattern. Screen printing and roller printing are the two most common printing methods used in textiles. Ink jet printing, heat transfer printing, and methods based on reliefs are also practiced in textiles.

筛网印花/滚筒印花

喷墨印花/热转移印花/凸纹

8.2.3.1 Direct Printing

直接印花

The most common approach for applying a color pattern is direct printing[see Fig. 8 – 5(a)]. The dyes are printed directly at the required places of the multi-colored design on the fabric, leaving the other portions white. Thus a multi-colored print on a white ground can be produced.

8.2.3.2 Discharge Printing

拔染印花

The whole cloth is dyed a solid color first; then the design on the roller is covered with a chemical, which, when it is applied to the cloth, discharges remove the color from it in those portions that correspond to the design on the roller. The background is left colored and the design is white[see Fig. 8 – 5(b)]. The same depth of color appears on both sides, because the colored portion was piece-dyed first. Likewise, the color in the background can be discharged if the background is printed with chemical so that the design is left colored. Usually, however, the background is darker than the design.

色泽深度

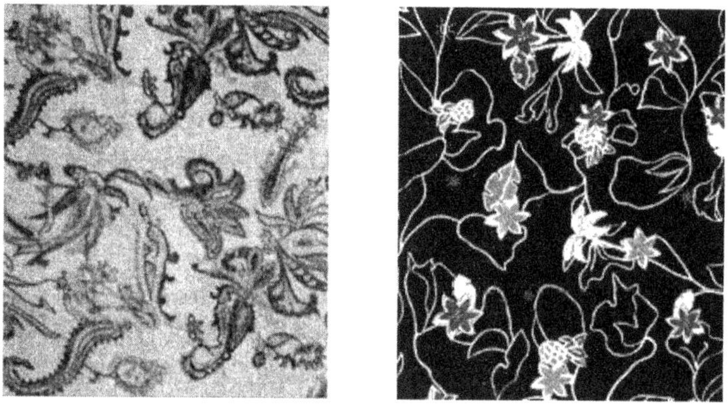

(a) Direct printing sample (b) Discharge printing sample

Fig. 8－5 Printing samples

8.2.3.3 Resist Printing

防染印花
防染浆
拔除

In this method, the design is printed first with a resist paste— a resinous substance that cannot be penetrated when the fabric is subsequently immersed in a dye. The dye will affect only the parts that are not covered by the resist paste. After the fabric has passed through a subsequent dyeing process, the resist paste is removed, leaving a pattern on a dark ground. In the discharge method, the fabric is first dyed and the color is then extracted by an imprinted chemical; in the resist method, a resist paste is first imprinted and the fabric is then dyed. The durability of the fabric is not affected by the resist method.

Broadly speaking, dyes used for printing are the same as those used for piece dyeing or yarn dyeing, except that dyestuffs may be thickened with starch, gum, or resin to prevent a color from bleeding or running outside the outline before it is dray. When they are dried, printed cloths are passed over hot rollers and then steamed so that the colors are set. Any excess dye is removed by a washing after the steaming.

8.2.3.4 Screen Printing

筛网印花
精细的底纹
光化学
阴影效应
印花筛框
亮光漆,搪瓷/印花浆
橡胶浆刀
刮浆器

When a design called for delicate shading, the process originally employed to produce the pattern was similar to that used for reproducing photographs in newspapers. Today a photochemical process reproduces the design exactly as it was painted. Elaborately shaded effects can be printed exactly like the original and reproduced many times. The fabric is first stretched on a padded table. A printing screen, made of silk, nylon, or metal stretched on a frame, is placed over the fabric. The parts of the pattern on the screen that are not to take the print are covered with enamel or certain paints to resist the printing paste. The printing paste is poured on the screen and pushed through the pattern portion with a wooden or rubber paddle called a squeegee. When one section of a pattern has been finished, the frame is moved to the next section, and so on until the entire length has been completed.

For screen printing, a continuous operation has now been mechanized, so that several yards can be printed every minute. The fabric moves along a table and the automatic application of the screens is electrically controlled. An automatic squeegee operates electronically. Mechanized printing reduces costs appreciably for large batches. Whenever high-quality fashion prints on exclusive dress goods 大批量/时髦服装 or intricate patterns and big repeats are requisites, screen printing is an important process.

Fabrics with large designs in limited quantities are frequently screen printed by hand. Rayon jerseys, crepes, and other dress fabrics, luncheon cloths, bed- 人造丝针织物/床单/浴帘 spreads, draperies, and shower curtains are often screen printed. Although screen printing is a slower and more expensive process than roller printing, the pattern repeats can be large-up to 80 inches.

8.2.3.5 Roller Printing 滚筒印花

Roller printing is the machine method of printing designs on cloth by engraved 印花滚筒 rollers(see Fig, 8 -6). Color deposited in the engraving is transferred to the fabric in the printing process.

A furnisher brush picks up print formulation from the color trough and transfers it to 给浆辊 the print roll. A doctor blade scrapes color from the smooth, unengraved portions of 刮刀 the print roller and leaves the formulation in the engraved portions. The print formulation in the engraved portions of the print roller is transferred to the fabric. The lint 铜刮刀 doctor cleans the print roll of lint and trash as well as color picked up by the print roll- 杂质 er from previously printed areas of the fabric. The cylinder supports the print blanket 承压滚筒 which provides a flexible surface which allows the fabric to be compressed into the engraved roll to accept the print formulation. The back grey prevents contamination of the 印花衬布 blanket by print formulation which may pass through the fabric.

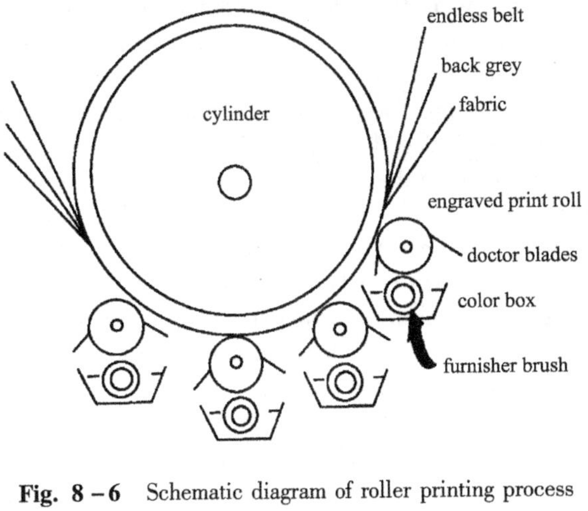

Fig. 8 -6 Schematic diagram of roller printing process

8.2.3.6 Ink Jet Printing

喷墨印花

This technique is used mainly to produce patterns on carpets. It is fundamentally different from any of the techniques described previously. Patterns are produced on the carpet by impinging very small droplets of dye formulation on the carpet in a predetermined pattern. Two basic methods are used. In one method, droplets of color produced by nozzles that open and close quickly on command from another method, a stream of dye flows from a color bar and breaks into droplets as the stream falls toward the carpet passing underneath. Air jets mounted perpendicular to the stream of color deflect the color into a catch pan when printing is not desired. The air jets are controlled by computer so that a predetermined pattern of color droplets can be deposited on the carpet surface. Ink jet printing requires a color bar or series of nozzles for each color in the pattern.

微滴 色柱 喷气嘴 推动杆

8.2.3.7 Heat Transfer Printing

热转移印花

In this technique the fabric is printed in nonpolluting color on large rolls of paper. The fabric in the greige and the printed paper are carried around a heated cylinder on a Nomex blanket. The heat transfers the design to the cloth. Heat, speed of rotation, and blanket tension are adjusted to the requirements of the cloth to ensure that virtually all the dye in the pattern is exhausted. The paper is then discarded as trash. Almost any design can be printed by this process.

坯布 诺梅克斯,聚间苯二甲酰间苯二胺纤维

Costs of production are kept down by storing fabric in the greige until needed to fill orders. Only the printed paper needs to be stored until required, at a much lower cost than storing fabric.

This method is suitable for printing on fabrics of the thermoplastic fibers, such as polyester, that melt at high temperature.

8.3 Finishing

整理

Textile finishing is a general term which usually refers to treatments on textile fabrics after dyeing or printing but before they are put on the market. The general aim of the textile finishing is to perfect the textile goods and render them fit for their end uses. For example, to improve the handle and appearance of the fabrics, stiffening, softening, or calendering etc., could be exerted on the fabric.

硬挺整理/柔软整理/轧光

On the basis of the technical feature the finishing processes can be divided into two main groups: mechanical and chemical finishes.

机械整理和化学整理

Mechanical process involve the passage of the textile material through machines whose mechanical action achieves the desired effects. This frequently accompanied by a heating process to enhance the desired effects.

Chemical processes may be described as those processes which involve the application of chemicals to the fabric. Chemical finishes are normally applied in the form of an aqueous solution or emulsion and may be applied via a variety of techniques, the main one being the pad mangle.

浸轧法

The distinction between the mechanical and the chemical finishing often fails, as in many mechanical processes chemicals are used or in many chemical processes the textiles are subjected by the mechanical action. Perhaps, these finishing processes should be called combined finishing. It means that both mechanical and chemical treatments are used to the textiles. The mechanical and chemical finishes used frequently are as follows.

8.3.1 Mechanical Finishes

机械整理

8.3.1.1 Calendering

轧光

Calendaring is essentially an ironing process that adds sheen to the fabric. The method varies according to the type of finish desired. Calenders are heavy machines made up of at least two rolls (see Fig. 8 – 7). One is usually of chilled steel; the other, of a softer material like wool paper, cotton fiber, corn husks, or combinations of cotton and corn husk. The rolls are supported in vertical frames. Plain rolling calendars may have as many as seven rolls—four steel rolls and three corn husk or cotton rolls. The steel rolls may be equipped to be heated by gas or steam. When the calender is in operation, the goods pass rapidly between the nips formed by the rolls. This is done at an average rate of 150 yards (135m) per minute.

熨烫/光泽
轧光机
冷淬钢
玉米的包皮
轧点

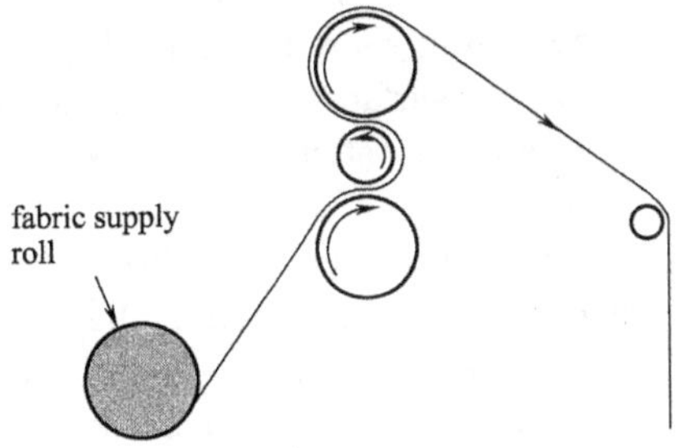

Fig. 8 – 7 Diagram of calender rolls, showing how fabric passes between and around the rolls

Wool fabrics are sometimes calendered or polished. The process adds a softer luster. Typical are wool broadcloth and sheen gabardine.

抛光
全毛绒面呢/光亮华达呢

A variation of the calendaring technique is the passage of polyester fabrics through a heat-transfer printing machine without the transfer print paper. The polyester fabrics acquire a soft, attractive sheen.

8.3.1.2 Tentering

拉幅

Tentering is the mechanical straightening of fabrics. It is a part of many finishing operations such as mercerizing, resin finishing, and drying. A clip or pin tenter or tenter frame may be used Fig. 8 – 8. The fabric is held horizontally by each selvage between clips or on pins. The chains holding the fabric gradually move apart to the desired width for the cloth before they enter the drying unit or other processes requiring tentered cloth. The tenter chains carry the fabric through the process and release the cloth to be coiled onto take-up rolls.

机械矫直
树脂整理/布铗或针板拉幅机
卷取辊

Fig. 8-8 Tentering of fabric

植绒 8.3.1.3 Flocking

Flocked fabric is made by gluing short fibers to the surface of a base fabric. Flock can be applied to the entire surface of the fabric to produce a raised surface or to only certain areas to produce a pattern. As shown in Fig. 8-9, flocking is done in a continuous process.

涂抹器 筛分 打浆辊 嵌入 静电植绒 红外线/介电加热

The adhesive is applied first, and then the fabric passes the flock applicator where the flock short fiber is sifted onto the fabric. Only a thin layer of adhesive is used so that the fabric does not become excessively stiff. Most of the flock must be perpendicular to the fabric surface in order to achieve the desired effect. In mechanical flocking, the fabric is vibrated by rotating beater bars beneath the fabric. This vibration helps the flock fibers which land vertically to become deeply imbedded in the adhesive. These vertical fibers then guide other fibers to land vertically. In electrostatic flocking, a charged field or air space is used to align the flock vertically. After the flock is applied, the adhesive is cured by convection, conduction, infrared, or dielectric heating to permanently bind the flock to the fabric.

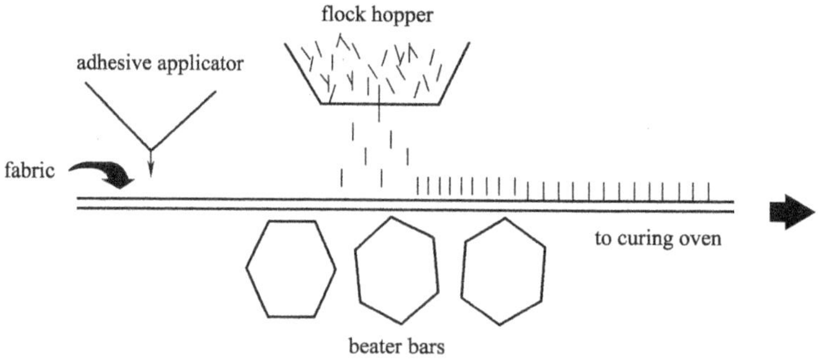

Fig. 8-9 Schematic diagram of a flocking process

8.3.2 Chemical Finishes 化学整理

8.3.2.1 Wrinkle-Resistant Finishes 防皱整理

A severe limitation of fabrics made of all or mostly cellulosic fibers is their tendency to wrinkle. Wrinkle-resistant finishes are applied to the fabric to make it resistant to wrinkling and to help it recover rapidly from wrinkling after wear. These finishes are also sometimes referred to as crease-resistant finishes. The effectiveness of the finish depends on the fiber content of the fabric, the construction of the fabric, and the particular chemical formulation applied to it. 配方

These finishes are all based on resins or reactants that will combine chemically with the fiber through a process known as cross-linking, whereby adjacent molecular chains of cellulose in the fiber are linked or tied together to provide greater molecular rigidity and to prevent intermolecular slippage. 试剂 相邻的分子链 分子间滑移

The original chemical finish used was urea formaldehyde; then came melamine formaldehyde and a host of variations on both basic chemicals including formaldehyde. These had a distinct disadvantage of giving off a fish odor, particularly when wet and causing some people to have allergic reactions such as eye tearing and skin irritation. Such resins release formaldehyde which has come under suspicion as a carcinogen. However, formaldehyde release from wrinkle-resistant finished fabrics has been greatly reduced as chemists have developed improved finishes. There are now resins which release virtually no formaldehyde. 脲醛/三聚氰胺甲醛 鱼腥臭 过敏

8.3.2.2 Water Repellent and Waterproof Finishes 拒水和防水整理

A water repellent fabric is one that will resist absorption and penetration of water for a given period of time, depending upon the length of exposure and the force of the water. Eventually, however, water will penetrate even a water-repellent fabric.

These fabrics have qualities not found in truly waterproof fabrics. They are porous, which permits the body to breathe, and are therefore comfortable. The hand and appearance of the fabric is good and very much like the untreated fabric. 多孔的

Some important types of water-repellent finishes are wax emulsions, pyridinium compounds, N-methylol compounds, silicones, and fluorochemicals. These finishes may be used to impart water repellency to various natural and synthetic fibers. Fluorochemical finishes impart both water repellency and oil repellency to fabrics while the silicone and repellents based on hydrocarbon hydrophobes provide only water repellency. The performance of fluorocarbon repellents results from the very low surface energy they produce. 吡啶季铵盐化合物 硅酮/含氟化合物 拒油性 疏水的碳氢链 碳氟化合物

For a fabric to be truly waterproof, it must be completely sealed with a substance that is insoluble in water. Coating of the fabrics with a low surface energy film former such as polyvinyl chloride or rubber makes the material completely waterproof. 表面能/成膜/ PVC，聚氯乙烯

8.3.2.3 Stain Repellent Finishes 拒污整理

Stain causes the most problems in the care of fabrics, so finishes that prevent stains are a real aid to the consumer. Stain-repellent finishes are made from such compounds as silicones, fluorochemicals, waxlike derivatives, triazine com- 沾污 蜡状衍生物/三嗪化合物

浮在表面 自由能 水性和油性的

pounds, and pyridinium compounds. The presence of the finish prevents stains from penetrating into the fiber by developing a high degree of surface tension on the fabric; by producing a substrate on the fabric, so oil and water actually "float"; or by presenting a surface with a very low free-energy factor, which prevents reactions between the stain and the surface. These finishes, which are applied to a wide variety of fabrics for clothing and home furnishings, are effective against water-borne and oil-bornestains.

柔软整理

8.3.2.4 Softening Finishes

柔软剂 取代氨基化合物

Softening finishes improve the hand and drape of a fabric. Softeners are among the most common textile chemicals. They include a variety of products, such as oil, fat, and wax emulsions, soaps and synthetic detergents, substituted ammonium compounds, and silicone compounds. The silicone compounds produce relatively durable softening effects.

黏度

The hand imparted and the properties of a softener depend on the physical state of the softener, its chemical structure, and its behavior on the fabric. Low viscosity liquid softeners produce a flexible, silky hand while solid softeners provide lubrication but less pliability. Since hand is difficult to assess objectively, softener performance is usually based on human judgement.

抗菌整理

8.3.2.5 Antibacterial Finishes

真菌 消毒剂/氯/过氧 微生物 杀菌

Growth of bacteria and fungi in textile fabrics is usually undesirable. Laundering with hot water and use of disinfectants such as chlorine or peroxygen bleaches destroys many microorganisms in textile products. Antibacterial finishes impart a self sterilizing quality to a fabric.

季铵盐表面活性剂 含氯有机化合物/银 镁/锌

Many chemicals are effective antibacterial agents for fabrics used in environments where growth of microorganisms must be prevented. Quaternary ammonium surfactants and antibiotics function as antimicrobial agents. Many chlorinated organic compounds and organometallic compounds containing copper, silver, iron, manganese, or zinc also make textile materials resist growth of microorganisms. Chlorinated organic compounds are also effective in inhibiting attack of insects on natural fibers composed of protein and cellulose.

床垫布

The main use of antibacterial finishes is for textiles which are being handled continuous by a large number of people. These include hotels, hospitals, institutional homes and student hostels, where mattress ticking, blankets and pillows, carpets and upholstery all come into contact with a large number of different individuals.

阻燃整理

8.3.2.6 Flame Retardant Finishes

A limitation of many textile fibers, especially celluloses, is their tendency to burn readily. The aim of flame retardant finish is to impart flame resistance to textiles. The flame-retardant textiles are mainly used in some special fields, such as work clothes of firemen, work clothes of workers in steel plant, military textiles, curtain hanging, carpet, children's clothes, etc..

可燃性

Three general methods are available for producing cloth with improved flammability:

① Flame retardant chemical finishes may be applied to the surface of cloth.

② They may be added to the polymer at some step in the synthesis process prior to extrusion of the polymer into fibers.

③ Inherently flame-resistant fibers may be developed and converted into cloth.

Most textile flame retardants change or interrupt the normal thermal decomposition process of the polymer. They may function in the solid substrate to lower the decomposition temperature of the fiber and change the nature and amount of volatile combustibles formed, or they may function in the vapor phase. For example, compounds containing phosphorus are converted to acidic materials which catalyze the thermal decomposition of the polymer. The flame retardants containing bromine or chlorine also decompose when the polymer decomposes to form volatile combustibles. The halogen scapes with the volatile combustibles and acts in the vapor phase to inhibit flaming combustion. For the flame retardant chemical to be effective, it must be available at the point at which the polymer decomposes.

热分解
易燃物
含磷化合物
溴
氯
卤素

Review Questions 思考题

1. What are the purposes of scouring?
2. Which method of bleaching is preferable?
3. Describe one important preparatory finishing process for smooth-surfaced fabrics.
4. List types of the common used dyes.
5. Explain the difference between dyeing and printing.
6. What are pigments?
7. What does resist printing differ from discharge printing?
8. Give a brief description of the transfer printing.
9. Why are finishing processes necessary?
10. What is calendaring?
11. What is the difference between water-repellant and waterproof?
12. Discuss shrinkage, telling why it occurs? Whether it can be prevented entirely? What methods could be used to prevent it?
13. List some types of special finishes, and give a brief description of each type.
14. Compare durable-press to wash-and-wear, give the methods of producing each.
15. List three general methods for producing cloth with improved flammability.

Suggested Activities 实训题

1. Decorate a plain woven white fabric by means of printing or dyeing.
2. Visit the fabric department of a large retail store. Identify the name of finishes discussed in this chapter. Describe the fabrics' appearance and finish characteristics.

Chapter 9 Textile Testing and Care
纺织品测试与保养

纺织品测试

9.1 Textile Testing

When we want to know what fibers are used in a fabric, or what properties are provided by a fabric, we must test the fabric by physical or chemical methods of textile testing. Table 9 – 1 shows some testing subjects for textiles.

Table 9 – 1 Some Testing Subjects for Fiber, Yarn and Fabric

Fiber Testing	Yarn Testing	Fabric Testing
Visual Examination	Yarn Count	Cloth Construction
Burning Test	Yarn Evenness	Fabric Faults
Staining Test	Strength-Extension	Strength-Extension
Solubility Test	Twist	Abrasion Resistance
Microscopic Examination	Abrasion Resistance	Evaluation of Particular Properties of Fabric
	Flexural Rigidity	
	Electrostatic Charge	

测试方法

9.1.1 Test Methods

In most testing laboratories, tests for physical properties are conducted at standard temperature and humidity conditions of 20°C and 65% relative humidity.

标准测试方法

9.1.1.1 Standard Test Methods

Standard textile tests require a specially equipped laboratory and must be applied at a standard temperature and humidity if the results are to be strictly comparable to those from other test agencies. Several organizations develop and publish test methods to evaluate textiles. In the United States, the two major organizations are AATCC and Committee D-13 of ASTM.

美国纺织化学家和染色家协会/美国材料试验协会/国际标准化组织

The international organization that publishes test methods for evaluating textiles is the International Standards Organization (ISO). Both AATCC and ASTM participate in the work of ISO through membership in the American National Standards Institute (ANSI). The American Home Economics Association (AHEA) participates in the process through representation on the ANSI advisory commit-

美国国家标准学会/美国家政学协会

tee. Many AATCC and ASTM test methods have been adopted by ISO and published as international standards. Test methods from Germany (DIN standards), Great Britain (BSI standards), and France (AFNOR standards) standards are also used.

德国工业标准协会/英国标准学会/法国标准协会

9.1.1.2 Other Test Methods

其他测试方法

Many trade associations such as TAPPI and INDA, as well as individual companies, have developed test procedures that are used to evaluate specific properties in specific products.

美国纸与纸浆工业技术协会/美国非织造布工业协会

9.1.2 Fiber Testing

纤维测试

9.1.2.1 Visual Examination

目测

This is the method employed to check the luster, handle, color, fineness, length, strength, and elongation of fiber by the naked eye.

9.1.2.2 Burning Test

燃烧测试

A burning test should be carried out on about half a dozen threads twisted together and applied gradually to the flame and allowed to burn slowly. The burning behavior of major fibers are presented in Appendix B of electronic version.

9.1.2.3 Staining Test

着色测试

These tests take advantage of the fact that dyestuffs, and a few other chemicals not normally used for dyeing, stain different types of fibers differently.

9.1.2.4 Solubility or Chemical Test

溶解或化学测试

This kind of test is needed when an unknown is very dark in color, when two or more fibers are present in a blend (which makes a burning test inconclusive), when fibers have very little visible structure, and in general, as a conclusive cross-check to burning or stain tests or microscopic examination.

9.1.2.5 Microscopic Examination

显微镜检验

Certain fibers have peculiar characteristics when examined longitudinally under the microscope, e.g., the scales on the surface of wool and the natural twist on individual cotton fibers. Much more information can be obtained from the cross-section (profile). Scanning Electron Microscope (SEM) has become a popular instrument for research work.

扫描电镜

9.1.3 Yarn Testing

纱线测试

9.1.3.1 Yarn Count

纱线细度(或线密度)

Although several yarn count systems are commercially used for various yarns, the tex system is now introduced as the standard of yarn count.

纱线条干均匀性

9.1.3.2 Yarn Evenness

The surface irregularity of yarn is of particular importance in relation to processing properties, as even with a well maintained mean count the appearance of the fabric can sometimes be affected considerably. For this test, visual evaluation, and electronic methods are provided.

视觉评价

强力伸长

9.1.3.3 Strength-Extension

Particular importance is often attached to the tensile strength attained in a single or folded yarn of particular type and composition. Yarn extension also plays a considerable role in the processing of the yarn and in the end-use properties of the fabric produced.

短纤纱和长丝纱的捻度测试

9.1.3.4 Twist Testing of Spun and Filament Yarns

Often a laboratory is required to ascertain what twist has been imparted to the fibers converted into a yarn on the spinning machine or twisting machine. In the case of textured filament yarns, the amount of crimp contraction and the resistance of the crimp to relatively high load are important factors.

耐磨性

9.1.3.5 Abrasion Resistance

The surface finish of the yarn plays an important part in its processing, e.g., it determines the friction occurring on thread guides. A measurement of abrasion resistance is the number of cycles required to break the test specimen at a given initial tension.

给定初张力

抗挠刚度

9.1.3.6 Flexural Rigidity

Under some circumstances the flexural rigidity of a yarn is also important. Included here we may mention particularly yarn feed on knitting machines and sewing machines.

静电荷

9.1.3.7 Electrostatic Charge

The occurrence of electrostatic charge is very troublesome during processing, especially in warping and beaming. Attempts are made to effectively counteract these undesirable influences, which are particularly evident with synthetics.

织物测试

9.1.4 Fabric Testing

织物结构

9.1.4.1 Fabric Construction

The fabric properties required influence the construction technique to be used, the use of weaving or knitting machines appropriate for the existing purpose, and finally the finishing technique. The fabric construction includes weave or structure, fabric density and weight per square meter, and thickness and surface uniformity of the fabric.

9.1.4.2 Fabric Faults

织物疵点

Faults found in fabrics can have various causes, e. g., the result of the spinning, weaving or dyeing process. Therefore, based on experience and a knowledge of textiles, several kinds of physical and chemical tests are required.

9.1.4.3 Tensile Strength-load-extension Characteristics

拉伸强力—负荷—伸长特性

In this test, tensile strength, tearing strength, thread slippage, bagging and bursting and wrinkle test are included.

9.1.4.4 Abrasion Resistance Testing

耐磨测试

An abrasion test carried out under defined conditions is intended to give guidance on the expected performance of textile fabrics under abrasive wear in practical usage.

9.1.4.5 Colorfastness Testing and Ratings

色牢度测试和分级

Samples given any colorfastness tests can be rated for staining of white cloth or for color change. Scales of standard rating evaluation greatly help consistency and are not expensive.

Colorfastness is a term used for the degree to which a dye holds "fast" to the fiber or fabric. A high or good fastness means that the dye does not fade in light, bleed or run in washing, crock or rub off in wear. Colorfastness is also measured to perspiration, seawater, pool water (containing chlorine), dry cleaning, weathering, various types of washing, fume fading, as well as finishing processes such as milling, decating, and others.

9.1.4.6 Absorbency or Wicking

吸水性或芯吸效应

This can be demonstrated using an approximation of the Weirick test (Sears Roebuck). Cut identical strips of various fabrics about 25 cm × 2.5 cm; suspend over containers of water so that the strips will be under the surface about 2.5 cm. Enter all strips at the same time, and note the rise of water on each strip after one minute (easier to see if coloring is added) and after five minutes, and record when the water reaches the top of the strip, if it does so in 30 minutes.

9.1.4.7 Thermophysiological Comfort

生理热舒适性

Standard tests for thermophysiological comfort of clothing are being applied. Three measurable and significant factors are thermal insulation, permeability to vapor (perspiration), and wicking ability of fabric next to the skin.

绝热性/透气性

9.1.4.8 Protection from Ultraviolet (UV) Rays

防紫外线性能

Research goes on investigating whether and how textiles can protect skin against UV. A good deal of early research was done by a textile scientist from Australia, M. Pailthorpe, who equated fabric properties to penetration by light to develop an Ultraviolet Protection Factor (UPF) and relate this to a Sun Protective Factor (SPF) that can be assigned to clothing as well as sunscreen lotions.

紫外线防护指数/日光防晒指数

织物特殊性能的评价

9.1.4.9 Evaluation of Particular Properties of Fabrics

Depending on the intended use, other special material properties are of significance to the serviceability, i. e. performance in actual use. Especially amongst these are noted stiffness, handle, drape, waterproofness, dimensional stability under the influence of moisture and heat, and finally various tests for determining serviceability.

纺织品保养

9.2 Textile Care

Textile products are complex items composed of many different parts. The conflicting interactions of fiber, yarn, fabric, and wet processing within one fabric are often difficult to predict. When two or more fabrics are combined in a garment or window covering, predictions become even more difficult, learning to care for these complex products is a major part of any study of textile science.

Textile products must receive periodic care to retain their fresh appearance and prevent damage from insects, mildew, and other destructive organisms. Dry cleaning and laundering are used for apparel and many household products; some additional procedures are also useful.

纤维性能与保养

9.2.1 Fiber Properties and Care

The way a fabric behaves in care is based on the reaction of the fibers to heat, water, mechanical action, and chemicals. This discussion covers the first three as they are encountered in home care, and deals mainly with the major fibers we use.

织物保养中热的作用

9.2.1.1 Heat Action in Fabric Care

Heat is one of the main factors that must be considered in care of fabrics, in relation to fiber content. Cellulose fibers (such as cotton or rayon) are resistant to heat, and so can stand hot wash water, a hot clothes dryer setting, and even ironing at high settings. Nylon and polyester, being thermoplastic, would melt under a hot iron, but will withstand hot water or a hot (cotton setting) home dryer. However, commercial gas dryers could reach temperatures unsafe for nylon. Ironing or pressing any thermoplastic fiber must be done with care (that includes all major MF fibers except those of reconstituted cellulose, such as rayon).

Acrylic fabrics should be protected from hot drying, or they may become lifeless and feel somehow thinner, although they may not have been otherwise damaged.

Heat-sensitive fibers should not be exposed to heat, as in tumble drying, without great care; in the case of olefin (polypropylene), avoid use of a heated dryer at all.

Heat has a harshening and drying effect on wool, silk, leather and fur.

Overdrying fabrics such as cotton sheeting, which are not actually harmed by the heat, can still leave the material feeling harsh against the skin, and develops more pilling than if articles are taken out when they are just dry.

织物保养中水和机械的作用

9.2.1.2 Water and Mechanical Action in Fabric Care

Fiber strength, especially when wet, is significant in home care. Fibers that are weaker when wet need support and TLC (tender, loving care) when they are washed.

Foremost is wool, often made into a bulky article that can absorb a lot of water; the wet garment is heavy. Wool in any article needs careful handling in washing, even if it has an antifelt treatment, because, like our own hair, it is weak when wet. Silk also loses strength when wet, but it is much stronger than wool to begin with, and fabrics made of it are often light. 防毡缩处理

Standard rayon and acetate both lose a great deal of their strength when wet and have only fair or poor strength to start with, so care is needed. HWM rayon has better dry strength than standard rayon and loses less strength when wet.

Cotton and flax, are significantly stronger when wet—ready for lots of mechanical action in laundering. Lyocell fibers, which are stronger than cotton when dry and do not lose a great deal when wet, can stand laundering procedures well. Synthetics start with good to excellent strength (except for spandex, present in small amounts as a rule) and change strength very little when wet. These types are the best candidates for machine washing care at home. With a durable-press finish, the strong cellulosic fibers can also offer total easy care.

9.2.2 Care of Various Textiles 各种纺织品的保养

9.2.2.1 Silk 丝绸制品

Silk garments should be cleansed soon after they are worn, to avoid the tendering action of perspiration.

Silk articles respond well to hand laundering, if the color is fast and there is little fear of shrinkage; heavier (suiting) fabrics of silk may shrink out of fit, unless they have been preshrunk. In any case, dry cleaning may be preferable for these heavier fabrics, as they may pillless and they may keep their shape better if not softened by washing.

Washable silk (washer silk) is expected to be machine washable, but it is imperative that you treat it as silk, not cotton; in particular, do not use built detergents (heavy-duty), or yellowing may result and more color loss than with the milder conditions as outlined for wool. Color on washable silk is often not fast, so get it dry quickly, either by tumble drying or rolling in a towel and pressing at once. (Dry cleaning may be no solution for this silk, as dyes may be more fugitive in dry-cleaning solvent.) Fabric softener is recommended for "sand-washed" silk, to prevent stiffening. In common one of the main weaknesses in less-expensive silk garments aimed at the middle market is inadequate seam finishing, quite aside from any special care needed for the fiber or fabric: with such smooth yarns, an enclosed seam such as flat felled is necessary to prevent yarn slippage along any seam that takes stress in wear.

Cultivated silk fabric (smooth, silky) should be pressed when slightly damp, on the wrong side, using a dry (not steam) iron set at warm, or use a damp press cloth. Silk may water stain if too damp during pressing. Wild silk is better pressed when nearly dry.

If the silk article is printed with more than one color those colors are probably not fast. If you wash it, prevent colors from bleeding: First, remove excess moisture by rolling it in a towel immediately after washing, and then press dry at once. Do not leave it lying damp, or colors may "migrate" and bleed.

Dry cleaning is the preferred method of care for silk fabrics and products. If handled carefully, however, some silk fabrics can be hand-laundered. A mild soap or synthetic detergent in warm (not hot) water should be used for silk, and minimal handling is recommended. Thorough rinsing is required, and the best method for extracting the water is to roll the garment in a towel and then hang it in a cool place, out of the sun, to dry. Silk should be ironed or pressed with medium to low ironing temperatures, and steam is acceptable.

过氧化氢/过硼酸盐
除臭剂
止汗剂/氯化铝
防护汗垫

If silk requires bleaching, hydrogen peroxide or perborate bleach must be used, for chlorine may destroy silk, One problem with silk fabrics is that body perspiration tends to weaken the fibers and frequently will alter the color. Many deodorants and antiperspirants contain aluminum chloride, which damages silk. It is often advisable to wear protective dress shields.

9.2.2.2 Wool and Hair Fibers

羊毛和毛发纤维制品

Woolen and worsted fabrics are widely used throughout the world. Wool fabrics tailor well, press easily, and can be shaped to conform to the body. A problem with wool fabric is its tendency to shrink.

胱氨酸
裂解

Wool is rather demanding in terms of care. It can be dry-cleaned and pressed easily, but laundering is difficult because wool cannot tolerate agitation or temperature change. Chlorine bleaches cannot be used on wool, for they break the cystine linkage and cause the fiber to disintegrate. When bleaching is required, hydrogen peroxide or similar bleaching compounds can be used. Mild soaps and detergents are recommended.

Following wear, woven wool garments should be placed on hangers and brushed carefully. This removes surface dust and soil and permits wrinkles to hang out. Knitted sweaters should be aired and then folded and stored in drawers to prevent sagging and misshaping from hangers. Frequent clearing reduces insect damage; unclean wool garments and blankets should never be stored, for they attract both insects and fungi.

Wool and hair fibers must be treated as protein fibers, not like cotton, using washing conditions that would be safe for our own skin and hair. Avoid alkali and high temperatures to prevent harshening and yellowing: no heavy-duty (alkaline) detergents. NEVER chlorine bleach, water lukewarm rather than hot. Keeping these cautions in mind, wool articles are laundered either by machine washing (with or without machine drying) or by hand.

9.2.2.3 Linen

亚麻制品

Linen (fabric made from flax fiber) takes vigorous washing well. The natural resistance of flax to chemicals including detergents, bleaches, other laundering aids, and dry-cleaning solvents combine to provide a product that is easily cared for. Further, these properties, plus resistance to sunlight, inherent fiber strength, and resistance to aging, result in fabric with long life. The preferred method of care for linen fabrics is laundering; however, dry cleaning can be used satisfactorily, and this procedure may be desired if end-use products are too large for laundering equipment or require special pressing. Avoid chlorine bleach, because linen can be attacked by it sooner than cotton. In general, linen fabrics do not soil quickly and, unless stained, do not require bleaching. It is safer to use a mild bleach like perborate. Wash dark-colored linens separately, and do not overload the machine, or you may get lighter streaks in the fabric. Linen demands TLC mostly in ironing (you could never call it just pressing!). Linen articles to be ironed must be thoroughly, evenly damp, yet cannot be left long at room temperature without danger of mildew growth; put dampened articles in the freezer to hold them safely for ironing. Linen calls for a hot (top setting), dry iron, and must be ironed until completely dry, or wrinkles form as soon as it is moved. When selecting apparel items of linen, consumers prefer fabrics with crease-resistant finishes. These may need some modifications in care procedures, particularly if stains or discolorations have occurred, as bleaches may discolor these fabrics.

9.2.2.4 Cotton 棉制品

Care of cotton fabrics, in general, is not demanding, unless they are high-fashion items. Most fabrics of cotton can be laundered and dried in home laundering equipment. Most detergents have no negative effects on the fiber. However, some colors and some finishes may be damaged and even destroyed by some detergents and laundry aids. It is important, therefore, for consumers to read labels carefully and adhere to suggested or recommended care procedures. To keep colors bright, or dyes from bleeding, use a lower washing temperature; if you wash with cold water, you will not get as efficient cleaning, although it may not be critical if the article is not badly soiled. If cotton requires ironing, it may be ironed easily at medium to medium-high temperatures.

Pretreatment of soils will greatly help in removal, especially oily soil at collars and cuffs and any oil spots. Counter mildew smell with a germicide or a fragrant fabric softener dryer sheet.

Ironing of firm cotton fabrics with no Durable Press finish is the same as for linen. Durable Press finished cotton and blends of cotton and polyester can be ironed without dampening, dry, with a steam iron.

9.2.2.5 Rayon 黏胶纤维制品

Regular viscose rayon tends to stretch and may shrink during laundering. However, fabrics composed of improved rayon fibers can be laundered easily and safely. (改良黏胶纤维) Although the laundry treatment and handling are dependent on yarn and fabric construction, finish and color application, and fiber content, rayon is not damaged by detergents and other laundry aids such as starches, fabric softeners, and water softeners. When it is essential for appearance, rayons can be bleached. Hypochlorite (次氯酸盐) or peroxide (过氧化氢) bleaches can be used, but they must be properly diluted to prevent fiber damage.

Special rayons such as flame-retardant fibers may require special care. Labels should be provided and the directions should be followed carefully in order to maintain the flame-retardant properties.

锦纶制品

9.2.2.6 Nylon

快干/免烫

Nylon is easy to care for. The first drip-dry, no-iron clothing was made of nylon. Hot water does not harm the fiber, but wrinkles and mussiness are minimized by the use of warm or cold water. An iron that is too hot may melt the fiber, glaze the surface of (scorch) the fabric, or actually make a hole or void in the fabric.

涤纶制品

9.2.2.7 Polyester

The major advantages of polyester fiber are its ease of care and wrinkle resistance. Permanent-press or durable-press textiles made of polyester and polyester blends require little or no ironing to maintain their smoothness and built-in pleats or creases. Polyester garments pack and travel well because the fiber is so wrinkle-resistant.

Polyester usually is advertised as "the easy-care fiber". It can be machine washed and dried. As with most thermoplastic fibers, better performance is produced if care is taken to avoid wrinkling the fabric when it is hot.

腈纶、烯烃类纤维、氨纶

9.2.2.8 Acrylic, Olefine and Spandex

Fabrics made from acrylic fibers may be laundered or dry-cleaned. Tumble-drying acrylic knits improves their hand, and most of the high-bulk fibers are labeled to indicate that the fabrics are made fluffier by machine drying. Pilling of knit acrylic fabrics is a problem.

Olefin fibers are easy to maintain if careful attention is paid to lower heat settings in washing, drying, and ironing. Very hot water and high-temperature drying may shrink improperly processed fibers. Prompt attention to oily stains is necessary. Follow care label instructions carefully for the best performance.

Spandex fibers may be laundered if reasonable care is taken to avoid high temperatures and damage from chlorine bleach. Most spandex garments have care labels recommending that chlorine bleaches not be used. Nonchlorine bleaches are usually given as an alternative where bleaching is required. Spandex fibers may be dry-cleaned safely.

毛绒织物

9.2.2.9 Pile Fabrics

Pile fabrics need special care in pressing to prevent the raised (three-dimensional) surface from becoming flattened: this procedure applies not only to pile fabrics such as velvet, velveteen, and corduroy, but to lace with a raised pattern or heavily embroidered fabric. Press face down into a thick terry or velour towel; for velvet, use a special needleboard, or lay the fabric face down over a horizontal raised screen (for small areas, over a clean, firm, flat hairbrush); press with a steam iron. To remove wrinkles, a velvet garment may be hung in a steamy bathroom. Brush pile with self-fabric or a soft brush. Launder washable pile fabrics inside out, to avoid trapping lint in the pile: Remove from the dryer while still slightly damp, to prevent crushing.

丝绒毛巾

Carpet is a pile fabric and requires care different from most pile garment fabrics. Crushing and matting in use may be construed by consumers as wear, but as long as the carpet pile is not worn away, it is considered to have performed satisfactorily, according to a typical carpet guarantee.

9.2.3 Care Procedures

保养程序

9.2.3.1 Laundering

洗涤

Laundering uses a solvent, a detergent, and mechanical action to remove soil. The basic process involves prespotting to remove stains, washing to remove soil extraction to remove water, and drying. Some items require pressing or ironing to remove wrinkles and provide a crisp finish to the fabric. Laundering is performed in home and in professional service laundries.

9.2.3.2 Dry Cleaning

干洗

Dry cleaning is a commercial cleaning process in which organic solvents are used in a specially designed machine to remove soil from textile products.

What happens in a dry cleaning plant? Basically this: Clothes are cleaned in a liquid solvent other than water (hence the "dry"); in comparison with home laundering, such solvents remove oily and greasy soil better, take out some soils or stains that are a challenge to budge in water, minimize shrinkage from fiber swelling, and minimize disturbance of colors and finishes, as well as keeping the garment's original shape and tailoring details which will be disturbed more in water, requiring restoration in finishing beyond the skills of the consumer using home equipment.

9.2.3.3 Ultrasonic Cleaning

超声波清洗

Ultrasonic (high frequency sound) waves were worked with in the 1990s to clean garments by shaking soil loose in a very short time; because soil was so thoroughly removed, with a fluffing out of the fibers, it was said that little pressing was needed in comparison with traditional processes, "dry" or wet. The process was aimed to be efficient and yet good for the environment, but still has not made much impact in the commercial cleaning field.

9.2.4 Care Labeling

保养标签

Many countries have developed systems for labeling textile products to denote care, in some cases mandatory, in many others voluntary, but with a specific system available. In most, symbols are used to represent various aspects of cleaning and maintenance of fabrics, as a means of offering information without the use of language.

Care labels giving instructions by means of symbols are found on textile articles from many countries. These are all very similar to those in a system proposed by the ISO (International Organization for Standardization). This system (see Fig. 9-1) has been formulated (and is owned as a trademark) by Ginetex—the

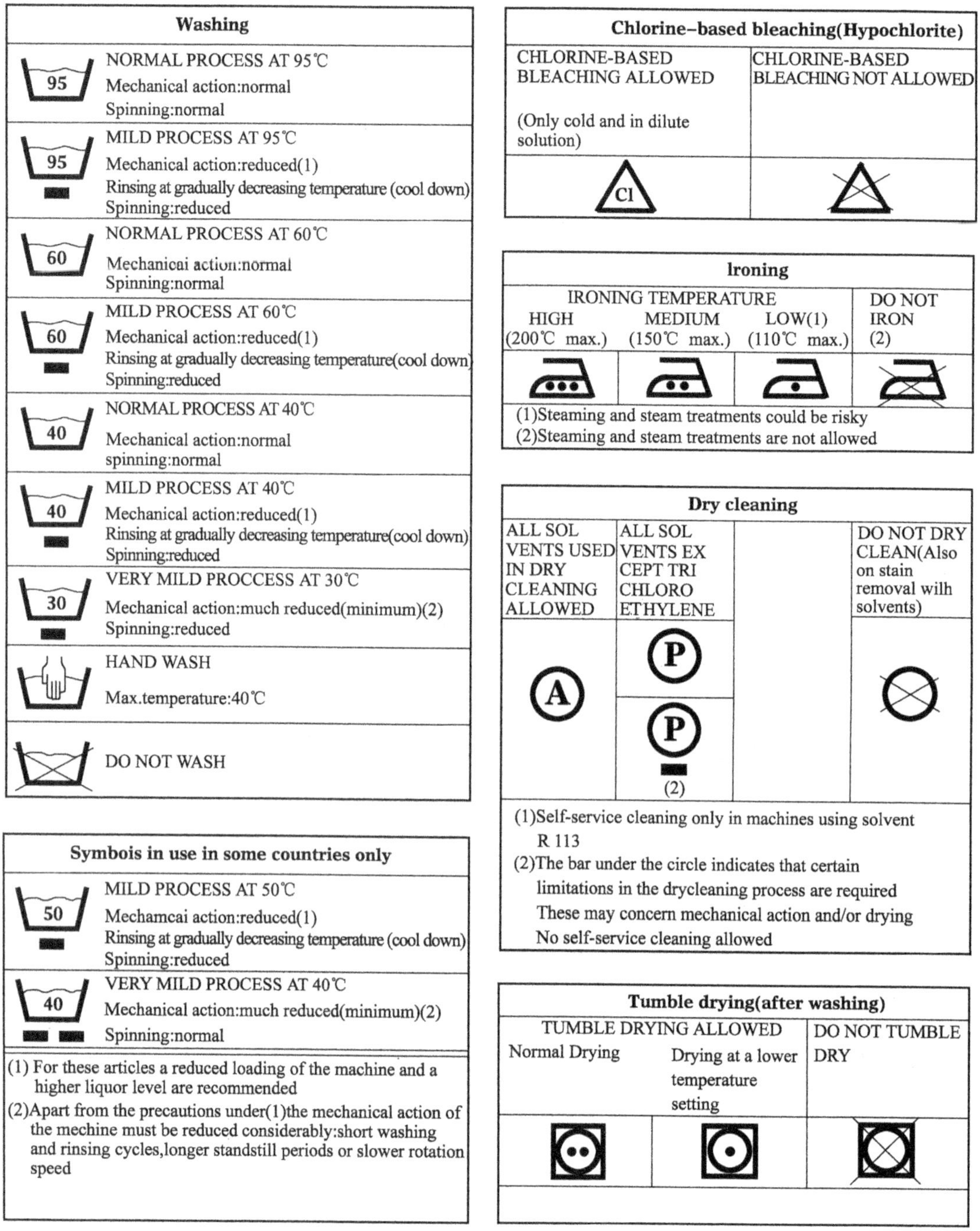

Fig. 9 – 1 International symbols for care of textiles labeling, registered to the nonprofit organization Ginetex

International Association for Textile Care Labelling, which licenses use of it to members. These symbols are included in ISO 3759: Care Labeling Code Using Symbols, published in 1991. The trademark logos used by ISO and Ginetex are shown in Fig. 9-2.

Symbols are printed in black only; a symbol with an X through it means: "Do not use this procedure".

(a)Trademark logo for ISO (b)Trademark logo for Ginetex

Fig. 9-2 Trademark logos

Review Questions 思考题

1. List some testing subjects for textiles.
2. What accounts for the similarity in burning of cotton, flax, rayon, and lyocell? Name two other fibers that would give similar results.
3. What can a microscopic examination reveal that a burning test cannot?
4. Outline the simple procedures that could be used to identify the following blends: polyester-cotton, rayon-wood, ramie-acrylic-polyester.
5. What are the main factors that must be considered in care of fabrics, in relation to fiber content?
6. What is dry cleaning? What advantages does dry cleaning over home laundering? What articles should be so cleaned? When does washing give a better result?

Suggested Activities 实训题

1. Obtain a washable fabric, and devise a procedure for testing it to determine color-fastness, dimensional stability, and appearance.
2. Obtain a sample of a fabric labeled "Machine washable" and a sample of a fabric labeled "Dry-clean only". Cut the fabrics into small squares and apply a stain to each sample. Mustard, grease, lipstick, fruit juice, grass stains, and ball-point pen ink stains are fairly easy to apply. Use some prespotting or stain removal techniques to remove the stains. Wash the samples labeled as washable to see how easily the stains can be removed.

References
参考文献

[1] Corbman, Bernard P. Textiles: Fiber to Fabric[M]. 6th ed. New York: McGraw-Hill, Inc. ,1983.

[2] David J Spencer. Knitting Technology [M]. 3rd ed. Woodhead Publishing Ltd. , 2001.

[3] Edward Miller. Textiles Properties and Behavior in Clothing Use[M]. London: Batsford Academic and Education Ltd. , 1984.

[4] Guang-wu Du, Fank Ko. Analysis of Multiaxial Warp-knit Preforms for Composite Reinforcement[J]. Composites Science and Technology , 1996, 56: 253 - 260.

[5] Horrocks A R, Anand S C. Handbook of Technical Textiles[M]. Cambridge, England: Woodhead Publishing, 2000.

[6] Isabel B Wingate, June F Mohler. Textile Fabrics and Their Selection[M]. 8th ed. Prentice-Hall Inc. TS 1449. W5, 1984.

[7] Ishida T. An Introduction to TextileTechnology[M]. Japan: Osaka Senken Ltd. , 1991.

[8] Jinlian Hu, Yaming Jiang. Modeling Formability of Multiaxial Warp Knitted Fabrics on a Hemisphere[J]. Composites: Part A, 2002, 33: 725 - 734.

[9] Joseph, Marjory L. Introductory to Textile Science[M]. 15th ed. Holt, Rinehart and Winston: The Dryden Press, 1986.

[10] Kathryn L Hatch. Textile Science[M]. USA: West Publishing, 1993.

[11] Kulkarni S V, Blackwell C D, Blackard A L, et al Textile Dyeing Operations[M]. Noyes Publications, Park Ridge, NJ, 1986.

[12] Lawrence C A. Fundamentals of Spun Yarn Technology[M]. London: CRC Press, 2003.

[13] Leong K H S, Ramakrishna Z M, Huang G A et al. The Potential of Knitting for Engineering Composites—a Review[J]. Composites: Part A, 2000, 31: 197 - 220.

[14] Lewin M, Sello S B. Handbook of Fiber Science and Technology[M]. New York: Volume Ⅱ-Functional Marcel Dekker, Finishes, Part A, B, Inc. , 1983.

[15] Lord P R, Mohamed M H. Weaving: Conversion of Yarn to Fabric[M]. Watford, England: Merrow Publishing Co. Ltd. , 1982.

[16] Lunenschloss J, W Albrecht. Non-woven Bonded Fabrics[M]. New York: John Wiley and Sons, 1985.

[17] Mary Humphries. Fabric Reference[M]. 3rd ed. New Jersey: Pearson Education, Inc. , 2004.

[18] Peter Schwartz, Trevor Rhodes, Mansour Mohamed. Fabric Forming Systems [M]. Noyes Publications TS1765. S38, 1982.

[19] Phyllis G, Tortora, Billie, et al. Understanding Textiles[M]. 5th ed. New Jersey: Prentice-Hall, Inc. , 1997.

[20] Schindler W D, Hauser P J. Chemical Finishing of Textiles[M]. Cambridge, England: CRC Press, Woodhead Publishing Ltd. , 2004.

[21] Smith, Betty F, Ira Block. Textiles in Perspective[M]. Englewood Cliffs, New Jersey: Prentice-Hall, Inc. , 1982.

[22] Trotman E R. Dyeing and Chemical Technology of Textiel Fibers[M]. 6th ed. New York: John Wiley and Sons, 1984.

[23] Walter Fung, Mike Hardcastle. Textiles in Automotive Engineering[M]. Cambridge, England: Woodhead Publishing, 2001.
[24] Warner, Steven B. Fiber Science[M]. New Jersey: Prentice Hall, Inc., Upper Saddle River, 1995.
[25] 島崎 恒藏 ほか. 衣服材料の科学[M].2 版. 东京:建帛社,2005.

纺织高等教育"十一五"部委级规划教材

书　名	主　编
纺织导论	薛　元
纺织导论(双语)	顾　平
纺织英语(第三版)	黄　故
新型纺纱(第二版)	谢春萍
纺纱实验教程	杨锁廷
机织实验教程	朱苏康
机织物组织与设计(双语)	聂建斌
织物纹织学	金子敏
针织学(第三版)(双语)	宋广礼　等译
针织工艺概论(第二版)	赵展谊
针织实验教程	张佩华
针织工艺与设备(第二版)	龙海如
针织厂设计(第二版)	李　津
针织 CAD 原理与应用	蒋高明
针织面料 CAD	龙海如
毛衫设计与市场开发	毛莉莉
纺织品功能性设计	吴　坚
纺织品及服装外贸(第二版)	张神勇
进出口纺织品检验检疫实务	郭晓玲
非织造布技术概论	马建伟
技术纺织品	陈韶娟
纺织企业现代管理(第二版)	林子务
棉纺织工厂设计(第二版)	钱鸿彬
纺织品检验学(第二版)	蒋耀兴
纺纱机械(第二版)	毛立民

书　名	主　编
织造机械(第二版)	陈　革
纺织厂空气调节(第三版)	周亚素
针织产品设计	张佩华
丝针织生产技术与新产品开发	陈慰来
丝绸产品设计(第二版)	沈　干
纺织经济与纺织品贸易	高长春
纺织服装电子商务	汤兵勇
技术经济分析	郑建国
纺织服装外贸英语函电	刘　嵩
纺织服装外贸跟单实务	倪武帆
纺织服装外贸单证实务	倪武帆
纺织品服装市场调研与预测	刘国联
纺织品服装消费科学	戴晓群
纺织服装贸易概论	王建坤
纺织服装外贸案例分析	吴雄英
纺织品服装进出口操作指南	钱竞芳
纺织品服装质量控制与管理	王亚超　杨卫丰
纺织服装企业物流管理	王亚超
现代纺织企业管理(第二版)	孙明贵
纺织电测技术(第二版)	朱正锋

书名	作者	定价(元)
【普通高等教育“十一五”国家级规划教材】		
纺织材料学(双语)	刘　妍	38.00
纺织 CAD/CAM	祝双武	39.80
【普通高等教育“十五”国家级规划教材】		
纺织材料学	于伟东	48.00
纺织品设计学	李栋高	46.00
机织学	朱苏康　等	38.00
纺纱学	杨锁廷	32.00
针织学	龙海如	32.00
【纺织高等教育“十一五”部委级规划教材】		
纺织品功能性设计	吴　坚　李　淳	40.00
针织学(第三版)(双语)	宋广礼　等译	40.00
纺纱实验教程	杨锁廷	38.00
机织实验教程	朱苏康	38.00
棉纺织工厂设计(第二版)	钱鸿彬	38.00
【纺织高等教育“十五”部委级规划教材】		
衣用纺织品学	蒋惠钧	35.00
纤维材料学	李栋高	48.00
纺织材料实验技术	余序芬	48.00
纺织品色彩设计	荆妙蕾	35.00
纺织概论	蒋耀兴	45.00
家用纺织品造型与结构设计	沈婷婷	31.00
纺织品 CAD 原理与应用	顾　平	40.00
纺纱系统与设备	郁崇文	58.00
纺织工艺设计与质量控制	郁崇文	42.00
纤维材料近代测试技术	潘志娟	48.00
针织服装设计与生产工艺	李　津	38.00
丝纺织工艺学	俞加林	56.00
羊毛衫设计与生产工艺	孟家光	45.00
【纺织高等教育教材】		
纺织面料设计	黄翠蓉　等	38.00
棉花检验与加工	徐　红	38.00
纺织服装营销学	王金泉	36.00
非织造布技术概论	马建伟　等	30.00
织物结构与设计(第三版)	蔡陛霞	34.00
纺织产品开发学(第二版)	滑钧凯	38.00
现代测试信号处理理论与实践	段　红	30.00
纺织英语(第二版)	黄　故	36.00

高校教材

注：若本书目中的价格与成书价格不同，则以成书价格为准。中国纺织出版社市场营销部门市、函购电话：(010)64168110 或登陆我们的网站查询最新书目。

中国纺织出版社网址：www.c-textilep.com